建筑·运算·应用 教学与研究 I

Architecture · Algorithms · Applications
Education and Research I

江苏高校品牌专业建设工程资助项目

国家自然科学基金项目（项目编号：51178116、51408172）

李飚 华好 唐芃 李力 编著

中国建筑工业出版社

图书在版编目(CIP)数据

建筑·运算·应用：教学与研究 Ⅰ / 李飚等编著.
北京：中国建筑工业出版社，2017.10
ISBN 978-7-112-21250-7

Ⅰ.①建⋯ Ⅱ.①李⋯ Ⅲ.①建筑设计-计算方法-文集 Ⅳ.①TU210-53

中国版本图书馆CIP数据核字（2017）第230673号

本书基于东南大学建筑学院建筑运算与应用研究所多年来的教学与研究成果，针对建筑生成设计、数控建造、物理计算等为代表的建筑学学科分支，梳理目前建筑数字技术的研究进展，探索建筑数字技术的核心价值与发展方向，及其对传统建筑学教育和建筑实践的潜在影响。

责任编辑：陈　桦　张　健
责任校对：李美娜　张　颖

建筑·运算·应用：教学与研究 Ⅰ
Architecture · Algorithms · Applications:
Education and Research Ⅰ
李飚　华好　唐芃　李力　编著

*

中国建筑工业出版社出版、发行（北京海淀三里河路9号）
各地新华书店、建筑书店经销
北京方舟正佳图文设计有限公司制版
北京雅昌艺术印刷有限公司

*

开本：787×1092毫米　1/16　印张：14½　字数：351 千字
2017年10月第一版　2017年10月第一次印刷
定价：**149.00元**
ISBN 978-7-112-21250-7
　　　（30882）

版权所有　翻印必究
如有印装质量问题，可寄本社退换
(邮政编码 100037)

东南大学建筑学院的前身是中央大学、南京工学院和东南大学建筑系。2003年，在原建筑系的基础上组建"建筑学院"。其是中国大学建筑教育中最早的一例，自1927年建系以来已走过90年历程。90年筚路蓝缕、成长壮大、传承创新，为国家培养了包括院士、大师、总师、院长等在内的大批杰出人才，贡献了大量重要的学术成果和设计创作成果，成为中国一流的建筑类人才培养、科学研究和设计创作的基地，并在国际建筑类学科具有重要影响力。值此90周年院庆之际，编辑出版《东南大学建筑学院90周年院庆系列丛书》，一为温故90年奋斗历程，缅怀前辈建业之伟；二为重温师生情怀和同窗之谊，并向历届师生校友汇报学院发展状况；三为答谢社会各界长期以来对东南大学建筑学院的关爱和支持。

　　这套丛书包括《东南大学建筑学院学科发展史料汇编1927-2017》、《东南大学建筑学院教师访谈录》、《东南大学建筑学院教师设计作品选1997-2017》、《东南大学建筑学院教师遗产保护作品选1927-2017》、《绿色建筑设计教程》、《建筑·运算·应用：教学与研究I》等共计6册。其中《东南大学建筑学院学科发展史料汇编1927-2017》完整展现了东南大学建筑学院各学科自1927年建系至今的发展历程，整理收录期间的部分档案资料，本书亦可作为研究中国近现代建筑教育源流及发展的参考资料；《东南大学建筑学院教师访谈录》收录了部分老教师的访谈文稿，是学院发展各阶段的参与者和见证者对东南建筑学派90年发展历程生动且真切的记录和展现；《东南大学建筑学院教师设计作品选1997-2017》汇集了近二十年来建筑学院在任教师的规划设计作品共计99项，集中反映了东南大学教师实践创作的成果、价值与贡献；《东南大学建筑学院教师遗产保护作品选1927-2017》依实践中涉及的建筑遗产保护五大类型，选有自20世纪20年代以来90余年完成的保护项目共65例；《绿色建筑设计教程》是近年来学院在建筑学前沿方向教改研究的成果之一，体现了在面对全球气候变化和能源环境危机时建筑学教育的思考与行动；《建筑·运算·应用：教学与研究I》着眼于计算机编程算法，在生成设计、数控建造和物理互动设计等方向，定义、协调或构建与城市设计、建筑设计、建造体系相关的各种技术探索，结合教学激发多样设计潜能。

期待这套丛书能成为与诸位方家分享经验的桥梁，也是激励在校师生不忘初心，继续努力前行的新起点。

编者识

序

东南大学建筑学院数字技术教学与科学研究起步于 20 世纪 80 年代，卫兆骥教授为此作出了开拓性贡献。1990 年，国家教育部批准东南大学"CAAD 实验室"计算机辅助建筑设计实验室为国家重点专业实验室。2000 年，联合国教科文组织在此设立"亚太地区技术网络服务中国中心"。2005 年起，相继开始探索生成设计等新技术及教学实验。2010 年，建筑学院设立"建筑运算与应用实验室"。2015 年，正式成立"建筑运算与应用研究所"，由李飚教授出任所长。经过三十余年的探索与发展，这一新兴的学科领域已超越传统的设计思维方法和一般性数字化工具的运用，在生成设计、数控建造和物理计算等方面取得了可喜成绩。此三大方向着眼于利用计算机编程算法，定义、协调或构建与城市设计、建筑设计、建造体系相关的各种技术探索，进而通过特定的算法技术，激发多样的设计创新潜能。

众所周知，数字技术正以前所未有的速度与建筑学科融合。建筑数字技术的迅猛发展极大地拓展了人们对建筑及其环境的认知方式及其相应的设计方法，对建筑的建造和相关产业产生了极大的影响，同时也对建筑教育提出了新的命题。数字建筑同样将建筑视为一种建造的技艺，并形成相应的建构文化。数字建筑的研究和实践具有鲜明的多学科联合交互的特性，它不仅涉及建筑学、计算机科学和电子信息科学自身的诸多问题，还涉及人文、经济、地理、气候、能源以及结构、材料、建造等多个领域的知识范围。这种跨学科的教学和研究不仅能帮助我们拓展视野，加强各相关学科之间的合作和交流，也促使我们重新思考和探索建筑学如何与各种新技术相融合以便产生更适应于人们需要的设计和更具可持续性的设计。可以说，运算已经成为推动建筑学学科创新的不可或缺的思维力量和技术要素。

2016 年，建筑运算与应用研究所主办以"建筑•运算•应用（Architecture • Algorithms & Applications）"为主题的国际会议和展览，意在对过去 6 年来的相关研究和教学实验进行总结，并邀约国内外该领域的部分专家学者展开探讨。这次活动引起与会学者的积极反响，并受到观展者热烈赞誉。本书收录了此次参展作品的主要内容，是对近年来建筑运算与应用研究所探索之路及其成果的客观记录。我认为，坚持严谨务实的学术作风，坚持以学科问题和社会需求为目标引领，坚持开放引进与自主探索的有机结合，坚持在学科交叉互动中实现技术创新，坚持人才培养与科学研究的融合发展，这是运算所在较短的历程中取得教学和研究突破性进展的基本信念基础。

建筑运算与应用研究所是学院最年轻的研究机构之一，拥有一支风华正茂、充满朝气的学术队伍。"回眸"显然不是当下的趣意，"展望"才更加吻合数字技术的本旨。与其他新型工程学科相比，建筑学科更需要充分意识到新需求和新技术的挑战，这也正是其创新发展的新机遇。建筑学科故有的人文创意属性则从另一个侧面提示了其数字技术发展的独特目标与路径。因此，我们亟需与相关学科和工程领域互联合作，以全新而专业的视角去学习和解读相关的学科知识、方法及其背后的思维特征。从这个视角看，生成设计、数控建造、物理计算可能是极具发展价值且需要深入拓展的跨学科研究领域，也是未来一个时期建筑运算与应用研究所的学术团队可以大展拳脚的领域。

东南大学建筑学院院长　韩冬青
2017 年 5 月 23 日于中大院

PREFACE

It was from the 1980s that School of Architecture Southeast University start the scientific research of digital technology with Prof. Wei Zhaoji's ground-breaking contribution. In 1990, the Ministry of Education approved the university's "CAAD Laboratory" for the National Key Professional Laboratory. In 2000, UNESCO established the "Asia-Pacific Regional Network for Technical Services China center" here. New technologies and teaching experiments such as generative design have been explored since 2005. In 2010, the School of Architecture set up "Laboratory of Architectural Algorithms & Applications" and the formal "Institute of Architectural Algorithms & Applications" (Inst. AAA) was established in 2015 with Prof. Li Biao as director. After more than 30 years of exploration and development, this emerging subject has gone beyond traditional design thinking methods and general digital tools with gratifying achievements being made in the generative design, digital fabrication and physical computing. These three directions focus on the use of computer programming algorithms for defining, coordinating and constructing various technical explorations related to urban design, architectural design and fabrication system, and then through specific algorithmic technologies to stimulate all kinds of design innovations.

It is well known that digital technology is being integrated with the field of architecture at an unprecedented speed. The rapid development of digital technology in architecture has greatly expanded the human consciousness about architecture and its environment as well as the corresponding design method, and greatly influenced the building construction and related industry. Meanwhile, it has also proposed new topics to architectural education. Digital architecture also considers architecture as a kind of construction method and eventually evolves to tectonic culture. The research and practice of digital architecture has a distinct characteristic of multidisciplinary integration. It not only involves issues of architecture, computer science and informatics, but also knowledges in respect of humanity, geography, energy, structure, material, construction and so on. Such interdisciplinary teaching and research can not only help us expand our vision and intensify the cooperation and communication among disciplines, but also promote us to re-consider and explore how architecture can get integrated with various new technologies so as to bring about more functional and sustainable designs. That is to say, computation has become an indispensable thinking and technical element to promote the innovation of architecture.

In 2016, Inst. AAA hosted an international conference and exhibition on the theme of "Architecture, Algorithms, Applications", which was intended to summarize the relevant research and teaching experiments over the past six years. Many experts and scholars in this field were invited from home and abroad. The exhibition aroused participants' positive responses and was warmly praised by audiences. This book contains the main contents of the exhibition works, which are the exploration and achievements of Inst. AAA in recent years. I believe adhering to the rigorous and pragmatic academic environment and leading by the academic and social function, combining open invitation and independent exploration, realizing technology innovation from interdisciplinary interaction and integrating personnel training and scientific research are the bases for Inst. AAA to get teaching and research breakthroughs in the past short time.

The Inst. AAA has a vibrant academic team, which makes it one of the youngest research institutes in our school. "Looking back" is clearly not its interest for the moment, "Looking out" is more consistent with the purpose of digital technology. Compared with other emerging engineering disciplines, the field of architecture needs to be fully aware of the challenges from new demands and technologies, which are also new opportunities for innovation and development. The humanistic creative attributes of architecture suggest the unique goals and path of the development of digital technology. Therefore, we need to cooperate with related disciplines and engineering fields with a new and professional perspective to learn and interpret knowledge, methodologies and thinking behind. From this view, the generative design, digital fabrication and physical computing might be highly valuable interdisciplinary research fields that need further exploration and future work from Inst. AAA.

Dean of School of Architecture, SEU
Prof. Dr. Han Dongqing
In Zhong DaYuan May 23rd, 2017

数字解读与技术递归

长期以来，建筑设计被置于理论和实践两面平行的镜子之间，相同的问题被无限复制，共同的原型被迭代递归，在"设计黑箱"的裹挟下数理运算缺乏直接展示其强大功能的机会。在过去的 20 年中，数字技术与建筑学学科迅速融合，其设计思想、设计方法、设计过程、建造流程、项目管理等方面都朝着更科学而系统的方向发展。复杂系统、人工智能与建筑数字技术相互嫁接逐步催生出以建筑生成设计、数控建造、互动设计等为代表的建筑学学科分支。一方面，建筑师对数字技术的解读不断变更，技术投入在实践与科研中获得了丰厚的回报，建筑数字技术正为建筑学学科充实崭新的理论与方法；另一方面，建筑数字技术本身也急需奠定系统的探索基础，数学与算法技术比以往任何时候都更符合建筑学的要求，并期待建立智能且包含众多接口的可扩展架构。

运算已经解决了科学领域的众多问题，但建筑运算的计算方法与它们大相径庭，其计算范式也不能被粗暴移植，寻求建筑晦涩定义的理性策略正成为学科共同的探索目标。对于建筑学的特定问题可以采取多种技术实现，但那些面向系统与框架的算法构建明显优于简单工具的等效替代。算法设计被确定并融入建筑数字技术研究，旨在寻求并获得算法设计的计算模式。特定的演化算法具有并行、进化和自适应特征，并将多种学科要素分解成直观的算法描述，甄别功能性因素与艺术性特征，以抽象的方式来获取最终产品的直接代理，进而提供全局优化的自适应解决策略，必将广泛应用于建筑学学科的模型建立。

与建筑数字技术相关的另一个重要角色是"创意代理人"：以数控设备为代表的机器工匠们。它们直接接触建筑材料，并在加入之初便以高效和精确为目标发挥其独特专长。在这个新奇的转置过程中，合理的空间与时尚的形式已经满足不了建筑师和业主的胃口，加工技术为建造提供了工厂组装模式，"数字链"系统一方面可以展示规则、非规则复杂要素的变化规则，另一方面它们仍包含成果与预定义的逻辑关联，同时祈求成果的多样性与个性化并存。

基于数字技术的设计方法为建筑设计提供了令人兴奋的机遇，它们并非传统设计方法的替代品，而是其有益的延伸。大数据的分析与提取正逐步形成统计分析和数据挖掘的数理与逻辑同构,并为建筑学学科提供科学的动态演化机制,形成彼此促进、互为依存的学科共生。建筑学学科已经到了一个拐点,算法与制造技术的成熟意味着建筑学正在进入一个前所未有的飞速发展时期,建筑数字技术的递归出口已经形成,各类匪夷所思正在变成现实,并呈现出目不暇接的视觉冲击和模式矩阵。

2017.10.

Interpretation of Digitization and Its Technical Recursion

For a long time, architectural design has been placed between parallel mirrors of theory and practice, which replicates the same questions infinitely and iterates universal prototypes recursively. Dissembled by the "black box" of design, the rational mathematical approach lacks opportunity to show its powerful function. During the past 20 years, owing to the rapid integration of digital technology and architectural discipline, its design methodology, construction process, project management and many other aspects are moving towards a more scientific and systematic direction. Complex system, artificial intelligence and architectural digital technology graft each other and help to emerge research branches of generative design, digital fabrication and interactive design. On the one hand, the architect's interpretation of digital technology is changing, and the technological investment obtains a lucrative return in practice and research. Digital technology is enriching architecture with new theory and methodology. On the other hand, architectural digital technology itself needs system-laying foundation for exploration. Mathematics and algorithm technologies are fitting the requirements for architecture more than ever and anticipating the establishment of intelligent and multi-plugin expandable structure.

Algorithms have solved many problems in science, but the methods of algorithm employed in architecture are very different and the calculation paradigms cannot be brutally transplanted. The rational strategies for seeking obscure architectural definitions are becoming common exploration targets. Specific architectural problems can be solved by a variety of techniques, but those algorithms facing system and framework are significantly advanced than the equivalent substitute of simple tools. The algorithmic design is identified and integrated into the research of architectural digital technology, aiming to obtain the computational model of the algorithmic design. Specific evolutionary algorithm has features of parallel, evolutionary and adaptive, and can decompose various disciplinary elements into intuitive algorithm description that distinguishes between functional reasons and artistic features. By the way of abstracting, it becomes the direct agents of final product and provides global optimization of the adaptive solution strategies, eventually being widely used in establishing the model for the architecture discipline.

Another important role associated with architectural digital technology is "creative agents": CNC machines, represented by equipment of computer numerical control. They are in direct contact with building materials and are uniquely focused on efficiency and precision from the very beginning. During this novel transition, neither functional space nor fashionable form would satisfy the appetite of architects and clients. Processing technology provides construction with factory assembly mode. "Digital chain" system can show the evolving rules of regular or irregular complex elements and meanwhile still contain the logic relationship between results and pre-definition, anticipating both diversity and personality from the results.

Digital technology-based design methods provide an exciting opportunity for architectural design, which are not a substitute for traditional design methods but rather a powerful extension. The analysis and extraction of big data are gradually forming the mathematical and logical isomorphism of statistical analysis and data mining and by providing scientific dynamic evolution mechanism for the architecture discipline, there forms a symbiosis relationship that is mutually promotive and interdependent to each other. Architecture has reached a turning point, with the maturity of algorithms and manufacturing technology marking that architecture is entering an unprecedented period of rapid development. The architectural digital technology's recursive exits have been formed and all the unbelievable things have become reality, presenting to us constant visual impacts and pattern matrixes.

2017.10.

目 录
CONTENTS

生成设计
Generative Design /1

01_ 音律柱
    Musical Column /8

02_ 映沙
    Sand Mapper /14

03_ 石头记（2016）
    The Stones (2016) /20

04_ 青奥村服务中心（中国·南京 2014）
    The Center of YOG Service Building
    (Nanjing,China,2014) /26

05_ 赋值际村
    Assign Ji /30
    生成设计思维模型与实现——以"赋值际村"为例
    Modeling and Realizing Generative Design:
    A Case Study of the Assignment of Ji Village /34

06_ 住区的生成
    Generative Residence /42

07_ 罗马火车站周边地区更新
    Renew Termini /48

08_ 体素建筑
    Voxel Architecture /54

09_ 泡与体
    Bubble and Volume /58

10_ 基于 RhinoScript 的视线控分析及其应用
    ——教敷营地块建筑设计视线分析为例
    The Sight Line analysis base on Rhino Script and Its Application
    ——Case study on the sight line analysis of Jiaofuying
    block /62

11_ 基于遗传算法的建筑体型优化
    ——以扬州南门遗址博物馆形体设计为例
    The Optimization of Architectural Shape Based on Genetic Algorithm
    ——Case study on the design of Yangzhou South City-Gate Ruins
    museum /66

数控建造
Digital Fabrication /73

01_ Angle-X /78
02_ Cell0046 /82
03_ Tri・V /86
04_ Dome・V /92
05_ Canopy /96
06_ Hakuna Matata /100
07_ Neuron /104
08_ Visual Robot /108
09_ 融・合（2014）
  Harmony・Peace (2014) /112
10_ 印象太湖石（2016）
  Taihu Stone Imagination (2016) /116
11_ 空影阑珊
  Sparse Shadow /122
12_ 龙舟记忆
  Dragon Boat Memory /126
13_ Ceiling Margin /132
14_ 槃
  Panzi /134

物理计算
Physical Computing /139

01_ 互动设计专题的最初尝试
  Original Attempt of Interactive Design /144
02_ 塑造景观
  Constructing Landscape /148
03_ 摆动的结构
  Wiggling Structure /152
04_ 媒体门
  Media Gate /158
05_ 动态立面
  Dynamic Facade /162
06_ 弦下
  Under the Sine /168
07_ 叶亭
  Leave Pavilion /174

08_ 动态交织
   Kinetic Weaving /180
09_ 信息墙
   Hexagon Info-Wall /184
10_ A Self-Organizing Wireless Sensor Network for Indoor Environment Surveillance /188
11_ Sequential Behavior Pattern Discovery with Frequent Episode Mining and Wireless Sensor Network /198
12_ 高精度多目标实时定位及分析系统
   High-Precision Multi-Targets Real-Time Locating and Analyzing System /210

后记
   POSTSCRIPT /214

生成设计
Generative Design

近年来，生成方法作为一种崭新的建筑设计手段逐步成为CAAD研究的重要分支，其独具匠心的系统模型也必将拓展建筑学方法。然而，可利用的参考资料、生成算法规则系统、人工生命系统、突现行为等等均起源于数学领域，它们提供了丰富多彩的实例和生成方法。通过对建筑元素的"自组织"优化组合，激发设计者借助传统方法不易获得的思想灵感，它是趋向艺术实践的程序创作系统，并将生成系统作为一种全新的生产方法。生成设计将计算机强大的储存及运行能力转化为对设计主体分析、探索的生成工具。计算机程序编写是该层面不可缺少的重要组成部分。建筑师运用计算机运算的高效性能；发掘建筑设计中可以交予计算机程序实现的部分；寻求并区分程序方法与自身经验的优化组合；深入计算机程序运算内核，探讨计算机程序相关算法等等。从而导出建筑设计的中间或最终成果框架。

建筑设计构思阶段、设计思路成熟之时，建筑形象已经基本分明，呼之欲出。相比之下，建筑生成设计技术所构思的通常为规则（如算法、约束等）制定，但根据规则而生成的结果则不可预料。透过多智能体生成技术基本方法，可以看到建筑设计计算机生成技术是一种科学的设计方法，建筑设计生成作品具有与众不同、不可重复的特征。它提供人类创作行为模仿自然的机会，同时也代表建筑设计方法的革新。计算机生成方法及其技术是跨学科的产物，随着相关学科研究成果的不断丰富，生成设计方法也须紧跟它们的研究步伐，从而真正做到从CAAd（drawing）到CAAD（Design）的革命性转变。

建筑设计基于"创造性"，并在"模糊性"中得到了充分的体现，生成设计模型应用于建筑设计方法需要建筑师从另一个角度审视建筑设计相关元素。将建筑设计系统理解为智能体交互协作组成的复杂适应系统，从而把建筑师的主体思维转化为建筑要素行为主体的建模过程。建筑生成设计模型将类型多样、数量巨大的建筑要素抽象为体现复杂系统特征的智能体集合，设计大多体现为各智能结构主体之间不断组合、分解的进化过程。各智能体"无意识"、"自私"的行为体现多智能体系统行为特征，多个建筑要素间相互作用表现出单个要素所不具备的总体特征，系统整体产生新特征的过程即为"涌现"，其整体表现优于个体的简单叠加，体现出"非线性"特征。建筑设计通常也被认为是一种程序，它类似于解决问题黑箱操作的方法，和创造力、革新息息相关。设计方法自身也正处于不断变迁的状态，随着信息科学的介入，革命性的理论、系统以及技术科技使得该过程新颖独特，不仅是它的产品，那些方法同样正在产生变革。就建筑学层面而言，建筑设计融入形式的视觉化过程及其各种元素的变形，并经常被反复推敲或者在一些候选方案中来回游弋，而擅于运用新技术的艺术家和设计师喜欢并知道根据新的方法改善他们的技巧套路。建筑设计生成技术整合建筑学、计算机科学、线性代数、计算机几何学、人工智能、复杂适应系统等等学科的技术及思维特征，并最终应用于建筑设计实践操作。

生成技术方法是一种科学的艺术创作过程，这种设计活动目的不仅仅要获得一个结果，更要形成一种可操作的程序编码，并作为一种技术手段进行操作和发展，逐步发展为可以利用的工具，也是一种具有强烈人文特征的设计方法，其步骤均始于设计者科学的思维或假设，并在设计之初便具有主观或者想象的未来形象预设。对于建筑师、设计师、艺术家而言，生成技术可以发展成为一种异常灵敏且有助于进行更为深刻的创造性设计的方法。生成技术是有计划的随机运作，确定性与非确定性、艺术与科学的统一，通过理性设计原则推导感性设计成果。引领建筑成为科学与艺术融合、理性与感性并存、人工与自然共生的客观产物。科学革命通常发生于传统与革新两种思维并存的时期，建筑设计生成方法呼吁思维定势的转换。"思维转变"与科学进步紧密相关，并影响于集体性的认知。新的理论和模型需要运用崭新的方法来理解传统的观念，并赋予新的学科内涵。

In recent years, generative design, as a brand new architectural design method, has gradually become an important branch in the CAAD research with its genuine system model surely to extend the methodology of architecture. The available references, generative algorithms, artificial life systems and emergent behaviors are all originated from the field of mathematics, which provide rich examples and generative approaches. Through the "self-organizing" optimal combination of architectural elements, generative design method will inspire designers in a way that the traditional methods aren't able to. It is a programmatic creative system inclining to the practice of art and makes the generative system a brand new production method. Generative design transforms computer's infinite storage and calculation power into generative tool that analyze and explore the design object, with computer programming as an indispensable part. Architects utilize the high efficiency of computation to excavate portions of the design that could be achieved by computer programs, seeking and differentiating the optimal combination of programming strategy and self-experience. They also look deep into the kernel of computer program operation and discuss the related algorithms in order to export the transitional or ultimate structure of architectural design.

In phase of conceptual design when the solution is mature, the architectural configuration is already clearly demarcated and vividly portrayed. However, generative architecture design focuses on the customary formula such as algorithms and restrictions, which will make the resultants generated from rules beyond anticipation. Via the basic strategy of multi-agent generative technology, we can see that the computer generative technology of architectural design is indeed a scientific design method, which yields unique and unrepeated production. It enables human creation to simulate nature and witnesses the reformation of architecture design. Computer generative methodology and techniques are interdisciplinary outcomes, so with the enrichment of the research of relevant disciplines, the generative methods must follow their steps in order to realize the revolutionary transformation from CAAd (drawing) to CAAD (design).

Originated from creativity and incarnated in ambiguity, applying models of generative design into architectural design method requires architects to examine relevant design elements in another perspective. Interpreting architectural design system as comprehensive adaptive system made of agent cooperation, architect's mindset is transferred to the modeling process oriented to architecture elements. Architectural generative design will abstract a big variety and quantity of architecture elements into agent congregation that features complex system characteristics, with design being in forms of the agent's main structure's constant combination and disaggregation process. Each agent's "unconscious" and "selfish" behaviors reveal the nature of multi-agent's system behavior characteristics and the interaction between multiple architecture elements reveals the overall feature that single element could not represent. The process of the overall system producing new feature is called "Emergence" which has superior behavior than simple add-up of elements, reflecting "non-linear" characteristic. Architectural design is generally reckoned as a sort of program, similar to "black box" operation for problem solving and is relevant to creativity and innovation. The design approach itself is going through the status of constant changes with the intervention of information technology, revolutionary theory and system as well as technology, which makes such process novelty not only in products but also in methods. In respect of architecture, the visualization process of integrating architectural design into forms and the transformation of elements are frequently examined or discussed among schemes. Those artists and designers who are good at utilizing new technologies know and are fond of revising their techniques based on new methodology. Architectural generative design integrates architecture, computer science, linear algebra, computer geometry, artificial intelligence, complex adaptive system and many other disciplines' technology and mindset to finally apply to the design practice.

Generative design is a scientific creation process of art, which aims not only for a result, but also to form an operable program coding and to operate and develop as a technical measure. Gradually it will develop into a usable tool and a design method with humanistic manner, whose steps originate from designer's scientific thinking or hypothesis and embody a future default in the beginning based on subjective or imaginative imagery. For architects, designers and artists, generative technology could be developed into a super sensitive design method that could enhance deeper creative design. Generative technology is a planned random operation which is a combination of certainty and possibility and unification of art and science. Via rational design principles, perceptual design outcome is deducted. Architecture is thus led to be the fusion of science and art, the coexistence of sense and sensibility and the objective product of human and nature's symbiotic relationship. Scientific revolution usually happens at the time when tradition and innovation coexist and architectural generative design calls for the transformation of stereotyped thinking pattern, which has a close relationship with scientific progress and will influence collective cognition. New theory and model need new approach to make new understanding of tradition and give it a new connotation.

01

音律柱 | Musical Column

设计者：2013 数控建造小组（本科四年级：郭梓峰，肖严航，吕一明等）| 指导老师：李飚
Designer: Digital Fabrication Group of 2013 (Grade 4: Guo Zifeng, Xiao Yanhang, Lv Yiming, et al.)
Tutor: Li Biao

"音律柱"是生成设计的练习之一，利用程序工具提取并分析音频数据，继而利用这些数据塑形。塑形过程采用分形原理，建立所提取数据与控制分形图案参数间的映射关系，产生连续变化的断面图形。一旦程序黑箱建立完成，仅需改变输入音频即可输出不同的生成结果。

"Musical Column" is an exercise of generative design that focuses on collecting datum from audio stream with programming tools and employ them for the shaping of columns. The shaping process adopts the principle of fractals for creating the mapping between the collected datum and the generated form, which generates constantly changing sectional geometry. Once the program is established, new configurations of columns can be generated by changing the musical input.

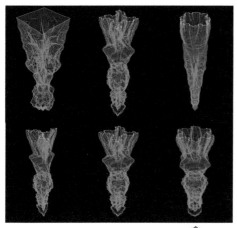
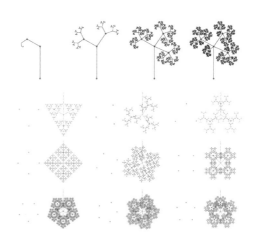

音律柱原理及生成图形（Generative graphics）

02

映沙 ｜ Sand Mapper

设计者：Inst. AAA 团队（2016）｜ 指导老师：李飚
Designer: Group of Inst. AAA (2016) | Tutor: Li Biao

"映沙"是一个以沙子为载体，以村落生成为主题的互动装置。沙子可由参与者任意调整堆砌，其形态经由 Kinect 传感器扫描并最终转换为地形数据。该地形数据随后被分析，产生若干地块，村镇聚落随即根据预设规则在各地块当中自动生成。

结合柔性模具的执行部分，展品由大量伸缩杆点阵排列，每个伸缩杆协同上下移动以点成面，拟合大尺度异形曲面的造型，也可应用于建筑曲面幕墙、雕塑等成型工艺。

Sand Mapper is an interactive device taking sand as the carrier with the theme of village generation. Sand can be piled arbitrarily by the participants, whose shape may be converted to terrain datum by a device named Kinect. The terrain datum is then analyzed to generate several plots suitable for the location of village while the villages are automatically generated according to rules predefined.

Combine with the operation device developed for points molder, which consists of matrixes of a large number of extendable rods. Each rod moves upward and downward synergistically to form a surface. It can form large-scale curved surface with the hundreds of actuators. This system is also designed for molding procedure of architectural curved curtainwalls, sculpture and etc.

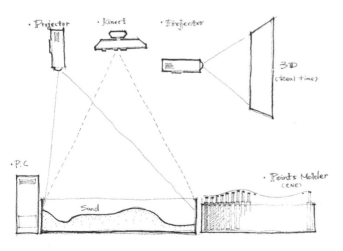

"映沙"与"柔性模具"原理示意图
(Schematic diagram of Sand Mapper and Point Molder)

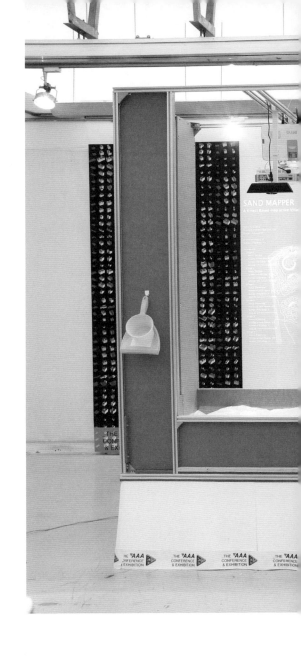

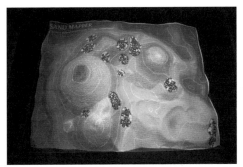

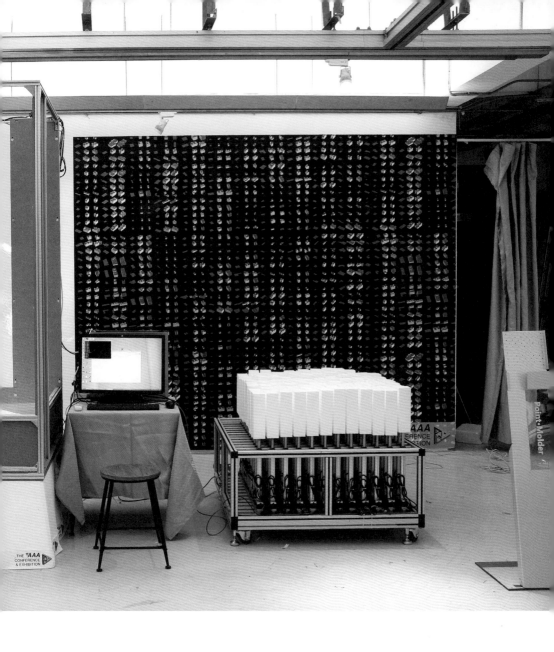

柔性模具（Points Molder）

石头记（2016） | The Stones (2016)

设计者：郭梓峰 | 指导老师：李飚
Designer: Guo Zifeng | Tutor: Li Biao

石头记基于拟合 Gyroid 极小曲面，以元球为辅助塑形手段，生成充满孔洞、造型奇特的酷似太湖石的形态。通过多次实验推敲，最终采用水平板片堆叠的形式。所有螺栓的位置、构件的切分位置及构件排版均由程序自动完成。一旦程序黑箱构建完成，仅需调整元球的数量、位置及半径，便可实现造型的快速修改。

表面通过三元表达式描述，每个点 p (X, Y, Z) 满足：

$$\cos(y)\sin(x) + \cos(z)\sin(y) + \cos(x)\sin(z) = 0$$

通过点阵拟合的方法模拟表面，可以获得满足上述公式的最小表面单元。然后，根据需要的表面外轮廓确定形状，最终，我们使用"元球算法"成功地获得预测的形状。

This project is based on the principles of *Gyroid* minimal surface and metaball, which define the surfaces as an auxiliary shaping method to produce geometries that similar to the Taihu Stone consisting of holes and irregular surfaces. The final geometries are realized through the layering of horizontal planes via multiple experiments. The positioning of bolts, the division of the components and the arrangement are all accomplished by the computer program automatically. Once the program is established, new results may be generated by changing the numbers, the positions or the radiuses of the metaballs.

The surface has a ternary expression with each point p (X, Y, Z) satisfies:

$$\cos(y)\sin(x) + \cos(z)\sin(y) + \cos(x)\sin(z) = 0$$

Minimal surface unit can be obtained by simulating the surface via the method of dot matrix. Then define the shape based on the required surface contour. Finally, we use the meta-sphere algorithm to successfully generate the shape predicted.

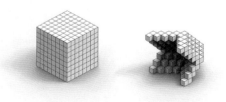

点阵　　　轮廓形状　　　结果

关键点　　靠近　　融合

一个三元方程定义
sin x·cos y+sin y·cos z+sin z·cos x=0

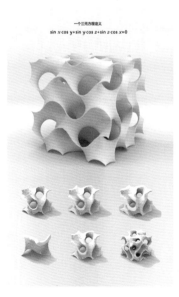

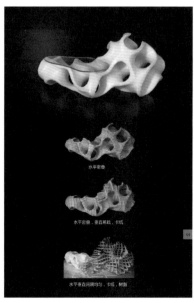

水平密叠

水平密叠，垂直稀疏，卡纸

水平垂直间隔均匀，卡纸，树脂

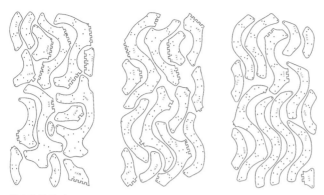

输出的构件（Output components）

构造细节（The details）

同一数据的不同塑形方法（Different fabricated ways based on the same data）

青奥村服务中心（中国·南京 2014）
The Center of YOG Service Building (Nanjing, China, 2014)

设计者：李飚，郭梓峰，季云竹 | 指导老师：李飚
Designer: Li Biao Guo Zifeng Ji Yunzhu | Tutor: Li Biao

青奥村服务中心为 2014 年南京青奥会所建，其建筑表皮设计为建筑运算与应用研究所和江苏生成设计事务所有限公司联合完成。生成原理为将图像数据采集并转换为构件的不同尺寸，该项目的处理程序可在数分钟内完成 7000 张构件全部加工图的绘制与导出。

建筑表皮材质选用银硅铝板，每一块板材由数量不等、角度各异的折板和直径不同的圆孔构成。借助折角角度的差异在天光的反射下形成深浅不一的明度，它们协同构成太空中地球轮廓外立面图案的矩阵。

The Center of YOG Service Building is designed for the 2014 Nanjing Youth Olympic Game. The façade design is accomplished by cooperative design effort between Inst. AAA and Jiangsu Generative Architectural Design Office CO., LTD. It employs the principle of collecting data from image pixels to shape the facade components. All 7000 detailed drawings can be generated within minutes using this project's operation program.

Silver silicon aluminum board is selected for the material of building facades, with each piece of the panel composited by different angled bended boards and different sized holes. With the different angles of the bended boards, it reflects diverse shades of sky light, and they collaboratively constitute the pattern of the Earth in the universe.

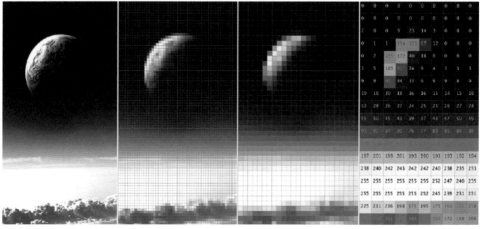

不同分辨率的图像灰度值提取（Extraction of gray value based on different resolution）

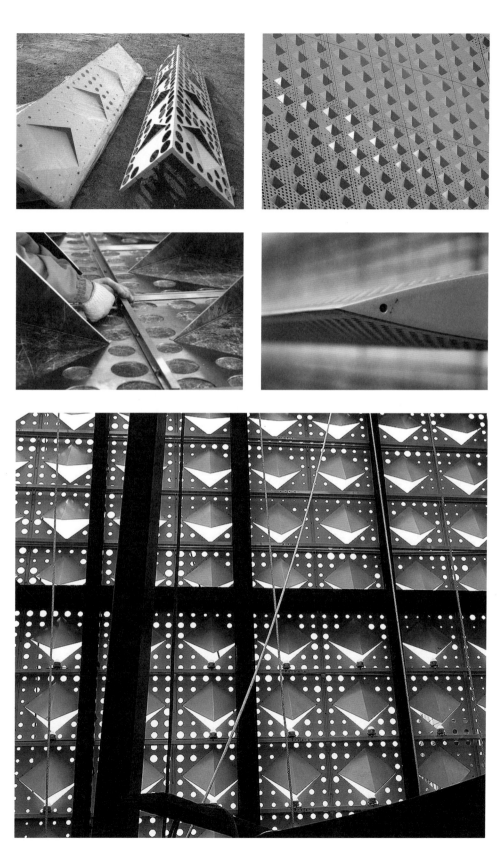

数控加工过程（Process of digital fabrication）

05

赋值际村 ｜ Assign Ji

设计者：2014 毕业设计小组（本科五年级：郭梓峰，季云竹，曹佳情，施天越）｜ 指导老师：李飚，华好
Designer: Graduation Project Group of 2014（Grade 5: Guo Zifeng, Ji Yunzhu, Cao Jiaqing, Shi Tianyue）
Tutor: Li Biao, Hua Hao

方案重点关注徽州村落际村的重建问题。它建立了一个徽州样式的民居生成体系，包括从村落的布局到建筑单体的深化设计。生成系统的研究重点在于村落的自组织系统以及对徽州民居的模式提炼。

This project mainly deals with the problems of the reconstruction of the Huizhou Village, Ji Cun. It realizes a Huizhou Style residence generating system, including both village planning and unit design. The main research focus of the generation system is the self-organizing mechanism of the village and the pattern extraction of the Huizhou Style residence.

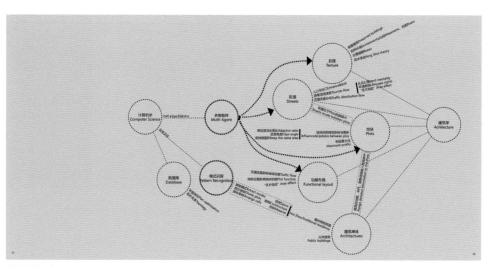

设计过程（Procedure）

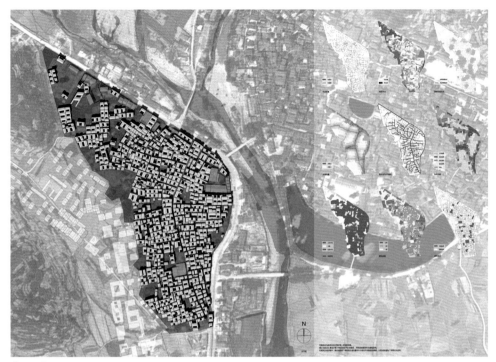

总平面图 (Site plan)

向量场模拟场地肌理 (Vector field as the orientations of houses)

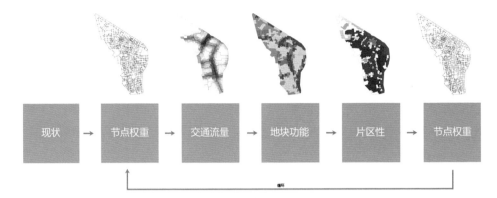

场地发展优化流程（Optimization process of the site）

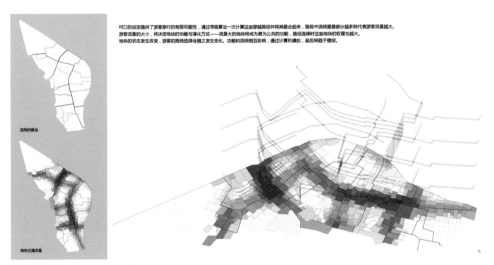

流线演算（Flow computation）

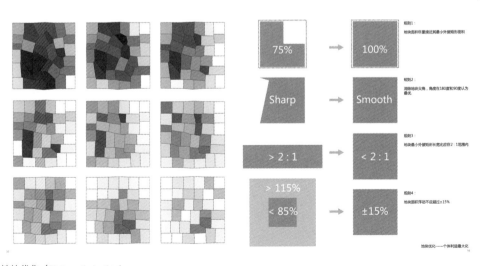

地块优化（Plots optimization）

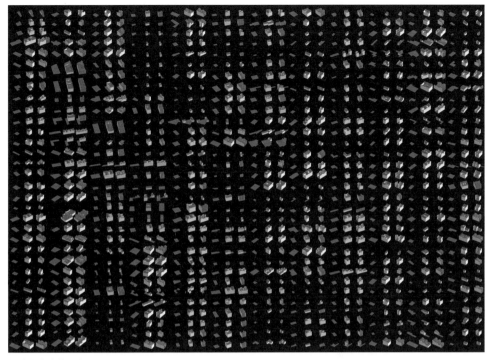

场地中生成的所有居民 (All of the generated houses in the site)

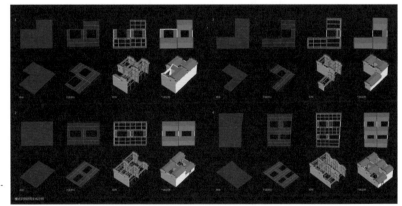

居民生成示例
(Examples of generated residences)

生成结果鸟瞰图
(Bird view of the generated village)

生成设计思维模型与实现
——以"赋值际村"为例
Modeling and Realizing Generative Design:
A Case Study of the Assignment of Ji Village

李 飚 郭梓峰 季云竹

摘 要：建筑生成设计提炼建筑原型，建立影响设计演化过程的时、空限定算法模型，进而运用编程工具动态优化预设的学科课题，实现演化模型所需的各种设计目标。文章以"赋值际村"为例，阐述理性模型、演化模型基本概念及其程序逻辑，探索际村的形态、肌理、交通、建筑功能和单体生成模式，寻求模型方法、程序手段与建筑现状的优化组合，为建筑学方法提供探索、实验的新思路。

关键词：模型 算法 演化 优化 生成设计

原文刊载于《建筑学报》2015 年第 5 期。

ABSTRACT: Architectural Generative Design extracts architectural prototypes to build time-restricted and space-restricted algorithmic models that affect the evolution processes of design. Further, the predefined problems are optimized dynamically by tools of programming to implement the desired goals of the evolution model. Taking "Assign Ji" as an example, the paper explains the basic concepts of the rational model and the evolution model as well as the logic of the related programming. Meanwhile, it explores the form, texture, transportation, building functions and building-generative patterns of "Ji Cun", and searches the best combination between models, procedural means and the existing conditions. It promotes a new thinking way for exploration and experiment of the Architecture Methodology.

KEYWORDS: model, algorithm, evolution, optimization, Generative Design

2014 年东南大学建筑学院照例参加全国近十所高校的联合毕业设计，这是一个以世界文化遗产宏村为背景的课题："建构——黟县际村村落改造与建筑设计"，基地位于安徽省黟县的际村（东西分别与水墨宏村、宏村相邻），总建设用地约 76000 余平方米（见图1）。学生根据调研结果自行选择设计目标，课题具有相当的开放性。东南大学建筑学院尝试划分一组学生以模型研究与程序算法为切入点，运用建筑数字技术探索古村

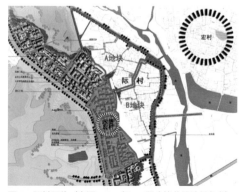

图1　际村区位（来源：2014 年 8+1 联合毕业设计任务书）

落肌理生成、建筑建构方式及与之相关的村落形态演变，取名"赋值际村"。

1. 理性模型：先验与有序的矛盾

程序设计师与艺术类设计师工作方式大相径庭，程序工程师的设计视角关乎实用和效益，兼顾效率和优雅；艺术家的设计则强调意义的传达和愉悦感。建筑师、产品设计师却需要二者兼顾。"赋值际村"课题不同于普通的例行程序设计，它不能借助改变参数逐个推演不同层面的设计对象以获得最终实现预设成果。理性模型也不完全适用于"赋值际村"，但可以为基本算法提供必要的思维流程，也是组员合作的抽象交流平台。

软件工程师对设计过程一般均有清晰但也隐晦的有序模型，假想用软件工程师的工作方式从事建筑设计，其工作流程大致如此：首先确定工程的目标及一系列"必要条件"，如：争取更佳的景观资源等；建筑功能、房间数量等；"效用函数"通常根据"必要条件"的重要性设定加权系数，所有必要条件线性相加组合，如窗户大小与节能的权重系数如何分配；对于只有满足和不满足的二元约束，接近边缘约束的权重代价会急剧增加。预算及资源分配是比较重要的指标，但在特定的项目中却未必成为设计权重的最大约束，如表征权力或特定文脉地段的建筑等。综合设计目标、必要条件、效用函数及关键预算，便形成理性模型的设计树，在每一个设计树的节点处会选择一条或多条路径，且树形结构组织的设计空间可以清晰描述，可以对诸多方案进行评估。理性模型类似对设计路径作穷举搜索，以搜寻最优解，选择设计树各节点上最有前途和吸引力的方案，如果路遇死胡同则会采用回溯的办法尝试另一条路径。这种理性模型概念是一个简单的线性过程，以可行性约束为依据，其思路也能被清晰描述。理性模型也可以在人工智能意义下搜索合适的标的，但模型比上述线性模型要复杂很多，对于设计过程自动化，人工智能仍是强大的先驱。

对于定义明确的项目，采用"先开始编码再说"的方法往往行之有效，但对于大型软件系统，没有系统化的理性模型规范则可能产生毁灭性的后果。理性模型会体现很多长处，必要的约束条件也有助于避免团队合作陷于不知所措的局面，后期可以有效规避大量的麻烦。但使用理性模型应对建筑设计问题时需要注意以下几个问题：

（1）理性模型越明确越有悖建筑设计多样解答的需求，建筑师的设计过程并非寻找一个真正的最优解，通常是寻求满足条件的众多解。

（2）通常在设计初期，建筑师只有模糊的、不完整的目标，定义不甚清晰的设计目标使得理性模型无从建立。建筑师通常基于既定的条件和规范，借助先验描绘出设计目标的大致轮廓，很少有设计师有机会或能力绘制完整的程序模型设计树。

（3）建筑师的设计树节点并不用于决策本身，而是一个临时方案。节点同时也可能映射到若干简单备选方案，通过树型结构模型所带来组合爆炸将是任何一台计算机无法承受。

（4）建筑师的先验直接判断往往是程序编写的噩梦，无论是直接或间接获得的先验均没有明确的设计树定义，更无法通过具体项目预测、跟踪下一个设计目标，而技术理性成功与否往往取决于专业人士对最终目标的认同。

（5）权重与约束的持续变化通常会导致判断条件也不断更新，例如，建筑使用者的特殊喜好会加重原先并未设置于设计树中的权重因子；建筑师围绕着各种变化着的约束做设计，并试图在设计空间中寻找创新和探索，这也是建筑设计的难点所在。

建筑师擅长空间思维，所以程序设计模式的可视化极有必要，设计过程的理性模型可以帮助并组织设计工作，有助于进行与项目相关的沟通。对于软件工程师则不必过于强调理性模型，因为这是他们专业与生俱来的经验和方式，但在具体操作时，线性的理性模型有可能具有巨大的误导性，有时它们并不真实反映工程师的真正工作流程，更不是设计师认同的设计本质。

2. 演化模型：模块与应用的分离

建筑生成设计研发不同于图形绘制或表现类的软件开发，演化模型的核心系统与建筑学效用模块彼此映成（一对多或多对一的映射），但核心系统的叠加并不能包络建筑学的全部。建筑学科与计算机学科思维方式及学科评价系统迥然不同，程序开发擅长运用理性模型处理定义明确的应用问题，不巧的是，大多建筑学问题却定义隐晦，这直接导致学科间的交流存在巨大鸿沟，所以至今生成设计研究尚限于以建筑学科人群为主导的有限人群，建筑生成设计的程序研发者同时也是软件工具的使用者，核心需求与模块实现以及程序运行反馈有效性的判断均由建筑学科人群完成。

演化模型基于对设计原型的递进式探索，它并不包罗"设计、构建、测试、部署、维护、扩展"等大型商业软件开发所必须的步骤。演化模型主要针对不能明确定义需求的开发目标，一般也不存在明确的优化收敛过程。演化模型可以提供大量原型后续研发空间。演化模型没有里程碑和合同的节点限制，开发模式可以采取分批循环开发的办法，每次循环便为原型的增添新功能，这种开发模式恰好对应建筑学科各类应用子集，如最优路径算法程序模块可以映射至疏散人流设计或城市设计的道路系统规划；模式识别算法可以映射至聚落模式及建筑形式生成，也可以用来探索各类城市形态研究课题[5]。演化模型基于设计过程中的问题空间和解空间的"共同演化"：问题空间和解空间共同演化，在此过程中信息在这两个空间之间流动[1]。演化模型程序开发需要将模块、程序数据结构与专业应用彼此分离，避免因子模块交叠而引起逻辑漂移的灾难性后果。

"赋值际村"基于设计组多次"试错式"程序探索，最终确定以演化模型作为程序开发基础。首先提取并分解该课题潜在的原型单元，如肌理形成内因、地块演化逻辑、交通系统与地块功能关联规则等等，一方面，它们基于建筑学科特定课题的应用需求，并将成为程序循环开发的功能模块预设；另一方面，分类方式需要结合开发者对程序算法与数理描述可能性的深度理解，并在确定分类的时候便可以借助理性模型大致描述。如：肌理与"张量场"概念存在一定关系，并借助于该数理方法模拟建筑群的生成肌理；地块演化逻辑可以基于半边数据结构（Half-Edge Data Structure）实现；交通系统可能与路径最优化算法相关，并由此拓展到地块的功能暗示；"模式识别"技术可以应用到中国传统建筑建构与形式生成探索等。换句话说，这种分类并不能根据单一学科确定，必须是开发者对总体知识结构（涵盖建筑学与计算机学科）而做出的理性预断。由于建筑生成设计成果的开发者与使用者合二为一，所以当工具核心需求实现后，研究小组可以通过不定期研讨提出有效的反馈，并在此过程中精化系统、增强系统能力，最终确定"赋值际村"的最终理性演化模型框架见2。

3. 赋值际村：模块与算法实现

际村与世界文化遗产宏村一河之隔，"赋值际村"基于村落的现状秩序，运用程序算法提取可程序化的建筑设计原型，从际村的形态、肌理、交通、建筑功能和单体模式探索其发展趋势，寻求模型方法、程序手段与建筑现状的优化组合。

3.1 肌理模拟

选择张量和场的数理方法作为控制村镇肌理生成规则之一是基于其数据图形化与城

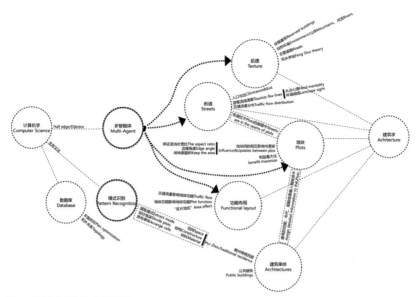

图 2 "赋值际村"演化模型框架

图 3 张量场形成的肌理暗示

市或乡村聚落肌理的形似，内在关联有待进一步研究，张量场均可以通过一组矩阵向量表示空间点的多重线性变换[3]。

"赋值际村"主要考虑四种关联因素：（1）河流、山体等既定位置关系，如，背山面水布局特征；（2）重要（保留）建筑，新建建筑与老建筑通常方向一致；（3）历史形成的主要干道，以及"鱼骨形"次一级干道，沿街建筑通常面向街道。（4）"风水"从另一个角度提供一些肌理控制规则，如祠堂的朝向等。有了以上四种控制因子，可以给各因子分配不同权重，同时判断何种分配权重最符合建筑学需求。图 3 显示不同权重系数对基地内矩阵空间的潜在肌理暗示。

肌理生成控制因子可以根据实际应用不断扩充，需要在程序编写时预留好必要的封装"接口"。肌理与各地块演化将共同影响建筑的主要朝向。

3.2 地块优化

地块优化体现为地块利益最大化演化目标的多智能体彼此博弈过程，地块需保持预定基本参数需求并遵循共同的演化规则，如：合理的长宽比、面积浮动范围、地块角度控制等等。"赋值际村"采用半边结构来描述地块与相邻地块的数据关系，相关地块内的建筑年龄、层数、功能及完好层度也被储存，以备整体演化之需。所有地块彼此无缝连接，道路网将通过半边结构节点与边的关系选择性在地块边界演化生成。

地块优化基于多智能体演化算法，并将规划与建筑控制参数植入编程代码，图4为随机初始状态下各地块的优化演变过程（地块灰度越深表示状态越差），各地块的演化遵守上述博弈规则。将其初始状态设置为际村既有地块数据，施用相同的演化规则便可以模拟所有地块动态优化进程。

多智能体系统演化模型结合半边程序数据结构可以动态解决复杂性和不确定性的规划与建筑学问题，单个智能体地块具有独立性和自主性，也能够基于预设的规则确定各自的优化方向，以自主推演的方式影响整体布局的优化进程。

3.3 街道与功能分区互动

"赋值际村"交通系统与地块优化互动生成，它们均基于半边数据结构系统，寻径算法已经很多，如A*、Dijkstra算法等。半边结构提供了顶点和边的基本数据（见图5-a），且可以根据现状路网设定不同的权重，这为街道生成提供了极大的便利。际村位于宏村和水墨宏村之间，考虑际村东西侧交通现状和空间关系，"赋值际村"在西侧和东北侧分别设置4个和6个出入口，见图5-b，两侧相通的路径共有24种可能，结合际村道路现状（见图5-c）便可以计算沿交通干道的地块，人流量多少也可以图像化显示，如图5-d（颜色越深人流量越大）。

临街且人流量大的地块更倾向于布置商业功能，街区内部的地块越倾向于居住功能，其他地块功能则介于两者之间。由此，可以得到一个动态优化的演化模型：地块自组织演化，不断寻找其更为理想的形状和位置，

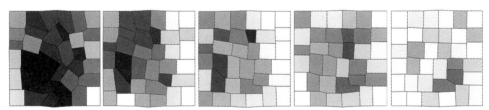

图4 地块优化演化过程

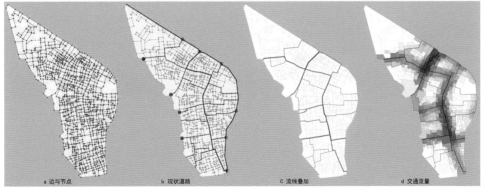

a 边与节点　　b 现状道路　　c 流线叠加　　d 交通流量

图5 "半边数据结构"道路网系统

并因此牵动主要交通网络的变更，变更的交通网反过来影响地块的功能属性，而地块功能的改变又会对地块形状提出新要求。地块、交通网络、功能互相牵制，当演化结果达到预设演化目标时，便可以结束整个优化演化过程，程序总体演化模型见图6a。此外，"赋值际村"尝试提供人为干预演化模型的可能，如果A、B地块被强制变更为商业功能，由于其规模较大，商业的集聚效应会影响其周边地块的功能，边与节点的权重将也随之改变。这种变化会成为新的演化条件，系统会根据新条件演化出不同的生成结果，如图6b。

3.4 模式识别与单体生成

传统徽州民居具有明显形制特征，在材料、做法、尺度、比例等均具有鲜明的地方特色，其外在表现通常与内部功能和建构方式息息相关。徽州民居的生成可以借助相应的"生形文法"规则完成。单体生成基于特定时刻的地块演化数据及总体肌理暗示（见3.1部分），"赋值际村"以徽派建筑作为其单体塑型模式。对于简单的地块程序经过预处理将其简化成四边形；复杂形状的地块则将其剖分为四边形子地块，程序评判并筛选出最佳的剖分结果（最适宜建筑建造的剖分，如图7），若出现得分并列则根据朝向和

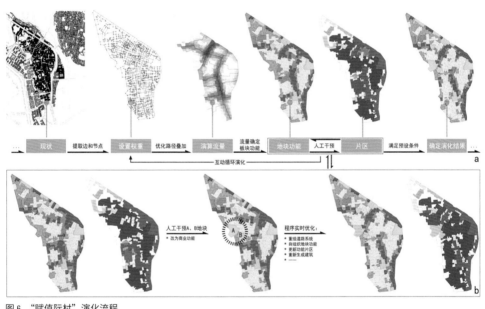

图6 "赋值际村"演化流程

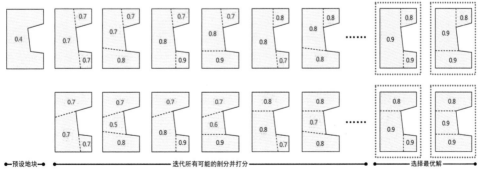

图7 地块剖分及评分模式

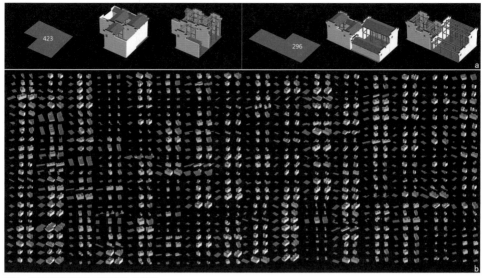

图 8　徽派建筑结构与表皮自生成系统

道路等因素做决定,程序将以最佳的剖分方式进行单体生成。

单体生成基于对徽派建筑的建构和风格的认知,是一个从内部结构到外部形式的语法生形过程,即地块剖分、木构架构建、徽派围护结构的模式识别。如图 8a 显示两例地块的徽派建筑自生成结果。图 8b 为其中 384 个地块的建筑生成结果,单体生成基于地块参数的线性推导,具有极高的生型效率和极强的自适应性,"赋值际村"只需要数秒钟便可以所有生成包含结构框架与围护结构的全部单体。

4. 成果整合与技术展望

上述程序模块遵循模型与应用相分离的原则,所实现的功能也相对独立,但在演化模型过程中它们均保持彼此数据的畅通交换,通过瞬间实现数十次之多的演化、评判、再演化的循环迭代,生成系统逐步提升演化成果的"设计"品质,其间,简单遗传算法起到关键性作用[6]。"赋值际村"源于对建筑学问题的理性演算与推导,但最终成果必须回归建筑学的专业评价。图 9 为演化系统的历时数秒钟的生成成果之一,对于不同预设参数和干预条件系统将生成不同的结果,并且程序每次运行结果均满足预设条件而显现成果各异。

"赋值际村"系统使用多智能体演化模型思想,定义智能体以及智能体各自独立的属性和方法,将建筑设计系统转化为由多智能体交互协作的复杂自适应系统,进而针对基于此生成系统的单元模式构建具有特殊适应性的"生形文法",从而形成完整的建筑生成设计的多模块、高内聚低耦合的徽州民居聚落空间生成系统,其模块及演化模型可以扩展到艺术、规划及建筑学等诸多学科。

演化模型系统在"赋值际村"中围绕直观的互动行为和抽象的思维逻辑而展开。并将宏观与微观有机联系,智能体与环境的信息互换使得个体的变化成为整个系统的演化基础,这种方式可以启发建筑师更多关于建模方式的思考——如何让程序提供更多科学的解答而非借助主观控制实现模式化的答案。对于复杂问题的求解,可以通过构建多

图9 演化成果（Results of generation）

智能体系统的方式，通过"自下而上"的"自组织"方式呈现不断趋于优化的结果。个体与个体之间的影响在自然界中无处不在，体现主体特征的个体与其所处的环境之间互相影响、互相作用成为系统演变与进化的主要动力。尽管模型方法可运用于个案各不相同，但技术方法却被应用于越来越多的相关领域。随着计算机程序算法技术的不断提升，各种建筑学经验和思维能力的系统模型与程序算法正朝着更广博、精深、复杂的研究领域扩展。众多以前无法处理的建筑学课题逐步找到新的计算机建模途径，其系统方法日益影响到各类学术分支，模式化的建模手段正逐步形成崭新的横断学科，其研究成果也加快了学科间的密切融合。

参考文献：

[1] Dorst, Cross: Creativity in the Design Process co Evolution of Problem-Solution, 2001.

[2] Frederick P. Brooks, Jr. 著，高博等 译：设计原本 [M]. 机械工业出版社 . 04/2013.

[3] Guoning Chen Gregory, Esch and etc.: Interactive Procedural Street Modeling, Proceedings. 2007.

[4] Ludger Hovestadt. Beyond the Grid: Digital Chain Reaction [M]. Basel: Birkhauser Verlag AG, 2010.

[5] Pascal Mueller, Peter Wonka and etc: Procedural Modeling of Buildings, ACM Trans. Graph., 62006.

[6] 李飚：建筑生成设计——基于复杂系统的建筑设计计算机生成方法研究 [M]. 东南大学出版社 . 10/2012.

[7] 李飚，韩冬青：建筑生成设计的技术理解及其前景，建筑学报 [J], 06/2011.

06

住区的生成 | Generative Residence

设计者：生成设计组（本科四年级）| 指导教师：李飚，华好
Designer: Groups of Generative Design (Grade4) | Tutor: Li Biao, HuaHao

该课题以居住类建筑为实践媒介，要求学生在掌握程序基本方法基础上，在算法研讨与程序实现相结合的实验方式中探索生成设计算法实现机制。理解程序算法规则在建筑学课题转化过程中的关键技术，合理提出适合程序运行规则的建筑学课题，敏感挖掘其程序实现机制，实现从"辅助绘图"（CAAding）到辅助设计（CAADesign）的思维转变。

案例1：Y住宅（2015）

设计者：唐浩铭 沈略

通过JAVA编程和对单元自组织行为的探索，该案例对景观资源、交通系统、居住空间进行算法整合，用计算机生成符合场地特性的住区方案。每个Y形居住单元具有三个方向的自组织延伸性，其所在的圆形映射住宅的容积率。

Taking residential building as an example of practice, this topic requires students to explore the mechanism of architectural generative design algorithms based on the combination of coding research and program realization. By understanding the key technology of the transition between program algorithmic rules and the architectural topics, architectural subjects suitable for the methodology of computer programming could be reasonably proposed, and the researchers need to sensitively excavate their program realization mechanisms and make the mindset transition from "CAAdraw" to "CAADesign".

Y Residence

Designer: Tang Haoming, Shen Lue

By means of JAVA computer programming and exploration of self-organizing behavior of unit, the project synthesizes the algorithms of landscape resource, transportation system and the living space to generate the residential layout appropriate to site features. Each Y-shaped living unit has three directions of self-organizing extensibility, where the circular mapping indicates residential FAR.

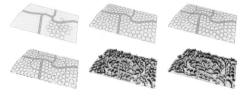

Y住宅（Y Residence）

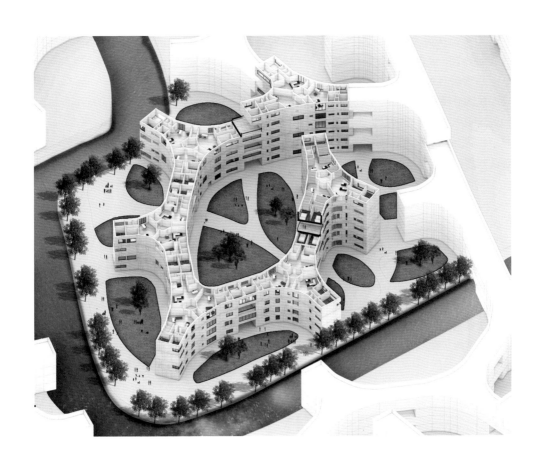

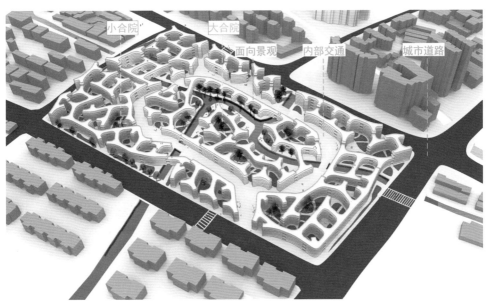

Y 住宅

案例2：住区邻里生成机（2016）

设计者：庞月婷 林欣 |

该案例借助算法探索邻里组团构成，以及建筑围合所形成的庭院交流空间与地块入口的逻辑关系，进而通过日照控制对户型进行自组织选择和删减，形成符合预设条件的住区规划。

案例3: WAVE DWELLING（2016）

设计者：梅琳丽 吴晓涵

用"冰裂纹"的生成模式对地块进行一次划分，若地块面积过大将进行二次划分，形成两个层级的路网。优化过程中赋予每个小地块边缘以张力，保证适当的长宽比，相邻地块间存在斥力，以便生成道路系统；根据地块面积，自行收缩膨胀，最终达到一个相对稳定的状态。

Residentail Neighborhood Generation Tool

Designer: Pang Yueting, Lin Xin

By means of computing algorithm, the project explores the structure of neighborhood and the logical relationship between the entrance of the block and courtyards formed by buildings. Moreover, the units of dwelling are selected and deleted automatically by program under the control of sunshine, in order to from the overall residential layout matching to the pre-defined condition.

WAVE DWELLING(2016)

Designer: Mei Linli, Wu Xiaohan

An algorithm for "ice crack" pattern is applied for the division of the main block and if the area of the block is too large, sub-division will be applied to form two grades of road network automatically. During the process of optimization, virtual tension is assigned to ensure proper aspect ratio for each plot, and repulsion exists between adjacent plots for generating road systems. Self-contraction or expansion is employed to reach a relatively stable status.

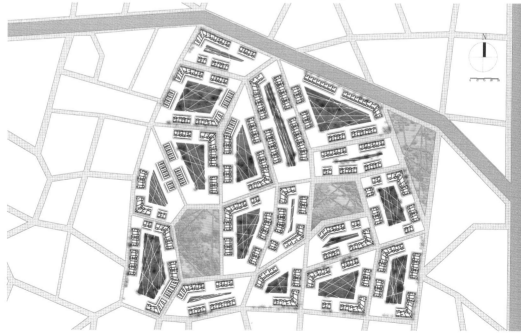

住区邻里生成机（Residential Neighborhood Generation Tool）

住区邻里生成机
(Residential Neighborhood Generation Tool)

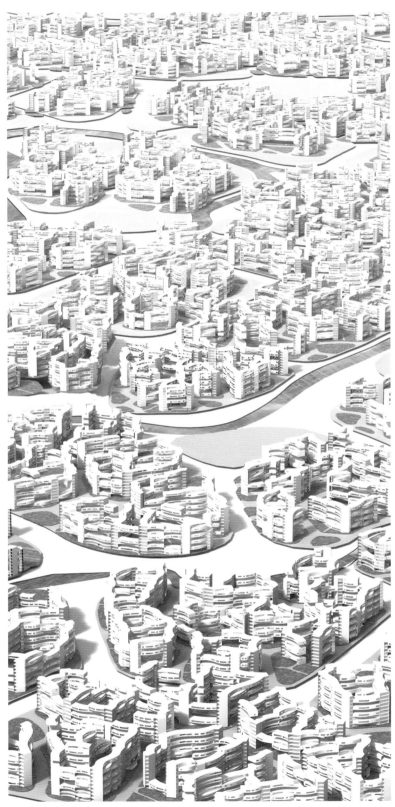
WAVE DWELLING

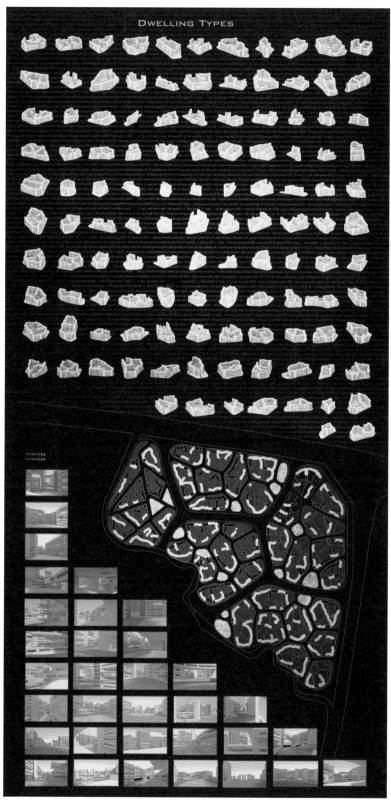

WAVE DWELLING

07

罗马火车站周边地区更新
Renew Termini

设计者：2016城市设计小组（研究生：李鸿渐，陈允元，徐佳楠，李思颖，韦柳熹，陈今子）
指导老师：李飚，唐芃

Designer: Urban Planning Design Group of 2016 (Graduate Students：Li Hongjian, Chen Yunyuan, Xu Jianan, Li Siying,Wei Liuxi,Chen Jinzi) | Tutor: Li Biao, Tang Peng

罗马特米尼中央火车站地区由于历史的发展，在罗马市中心形成了独特的城市肌理，其背后是多要素共同作用下的复杂系统。本小组旨在利用数字技术，通过网络共享的城市地图数据，将复杂的城市问题分层剖析。利用大数据的优势，建立城市空间的评价标准，提出可供参考的城市问题解决方案。

数据转译：通过对XML文件的分析，程序建立地块和周围的数据库信息，如地块的功能、高度、交通可达性等。据此设定类似待建地块的各类权重，并将其转译并植入新地块中。

交通弥合：根据城市现有交通系统确定重要的节点，并以此作为关键出入口，在火车站轨道上方创建一个慢行网状。

功能织补：改造景观系统，置入新的功能，通过连续的景观步道，将城墙以及遗址等景观点联系起来。

Due to the historic development of Rome's Termini Central Train Station, it forms a unique urban texture in the city center of Rome with a complex system. The research team aims to use digital technology and the urban map data shared on internet to analyze the complicated urban problems. We aim to employ the advantages of big data to establish evaluation criteria for current urban space and a reference solution for urban problems.

Data translation: Through the analysis of XML files, the program builds database of the plot and its context, such as function, building height and transportation accessibility, based on which they put different weights in such un-built plots and translates into new plots.

Traffic Bridging: Identifying important traffic nodes according to the city's current traffic system, and using them as key exits/entrances, our tool generates pedestrian network above the track of the train station.

Functions Knitting: By means of transforming the landscape system and inserting new features, city walls and historic heritages are linked by continuous pedestrian trails.

设计研究范围（The reseouch area）

基地现状（The site）

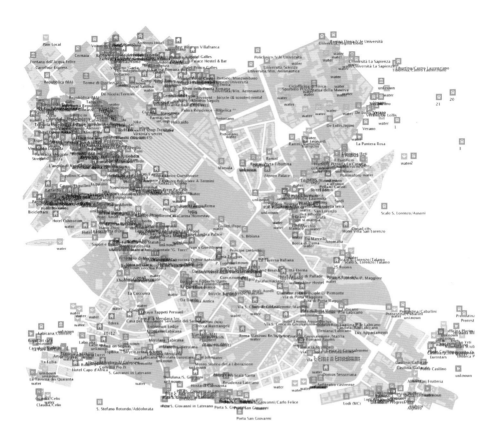

非机动车停放	公交站点	饮水点	历史性建筑
旅馆、酒店	医疗服务点	餐饮	购物商场

地图数据提取

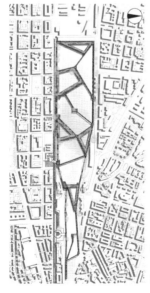

车站顶部慢行系统
(The pedestrian - system above station)

车站顶部慢行系统剖面图（Section of pedestrain - system）

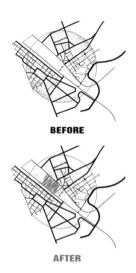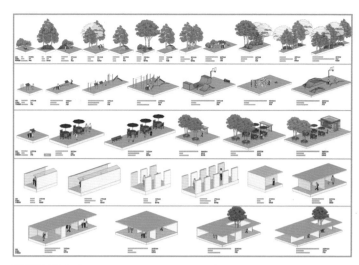

景观设计策略 (Strategies of landscape planning)

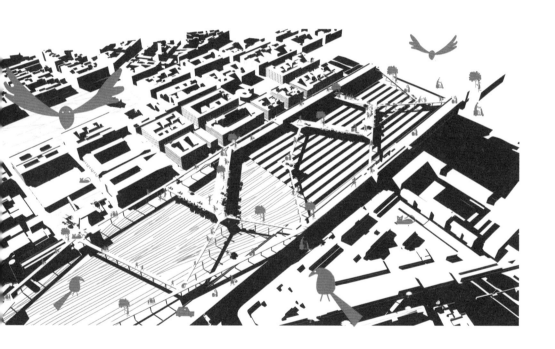

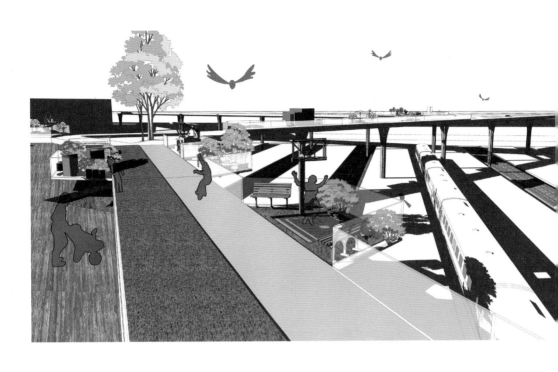

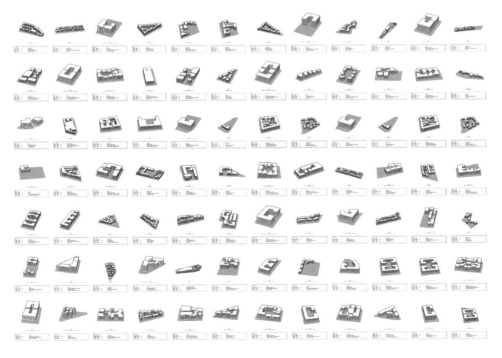

建筑形态数据提取

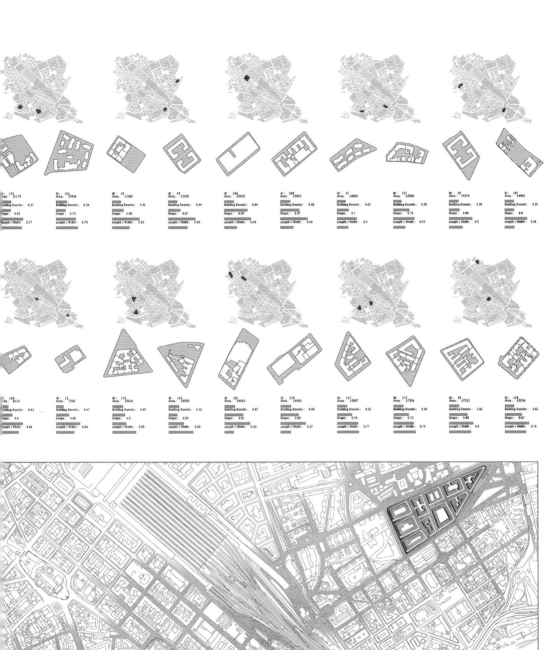

Termini 周边肌理织补（Sewing the texture around Termini station）

体素建筑 | Voxel Architecture

设计者：郭梓峰 | 指导老师：李飚
Designer: Guo Zifeng | Tutor: Li Biao

体素建筑受到自组织细胞的启发。基地被细分为体素（矩形网或 Voronoi 细胞），每个体素有两个状态，占用或为空。占用的体素构成了建筑物的体量。通过改变体素的状态，可以优化建筑物形体。

每个占用的体素都有一个对应的功能，如居住、商业或办公室。在优化过程中，体素的状态和功能均可以互换。不同功能在环境因素上有不同的需求，例如住宅需要更多的阳光，而商业需要可达性。实验涉及三个因素：照明条件，噪声水平和可达性。它们分别通过相应的评估函数进行评估。

设计人员规定占用体素的数量、不同功能的比例和不同的权重因子。优化采用进化策略，进化通过交换体素的状态实现，更好的改变被保留，糟糕的变异则被丢弃。优化实现了建筑体量及其内部功能组织的自动分布，结果通过过程建模的方法输出为网格模型。

体素建筑将目标由聚类改变为建筑学相关的标准。因此，交换操作不仅影响建筑物的形状，还影响其功能分布。体素在优化初始阶段随机分布，当优化完成时，它们被聚集成不同形状的组。

Voxel architecture is inspired by the self-organizing cells. Sites are subdivided into voxels (rectangular girds or Voronoi cells). Each voxel has two statuses, occupied or empty. Occupied voxels compose the volume of the building. By changing the status of voxels, the building can be optimized.

Every occupied voxel is assigned with a function such as residence, commerce or office. During the optimization, not only the status of voxels, but also the functions may be swapped. Different functions require different environment. For example, residence needs more sunlight while commerce needs to be easily accessed. In this experiment three factors, namely, lighting condition, noise level and accessibility, are involved. They are evaluated through corresponding evaluators respectively.

The amount of the occupied voxels, the proportions of different functions and the weights of different factors are specified by designers. The optimization adopts an evolutionary strategy that such evolution is made by swapping the status of voxels. Better changes are accepted and worse ones are discarded. The optimization achieves automatic distribution of building volume as well as its internal functions. The results are exported in mesh model using procedural modeling.

The voxel architecture changes the objective from clustering into architectural criteria. Thus the swap operation not only affects the building shape, but also the function distribution. Voxels are randomly distributed at the beginning of optimization and gathered into groups with different shapes when optimization is finished.

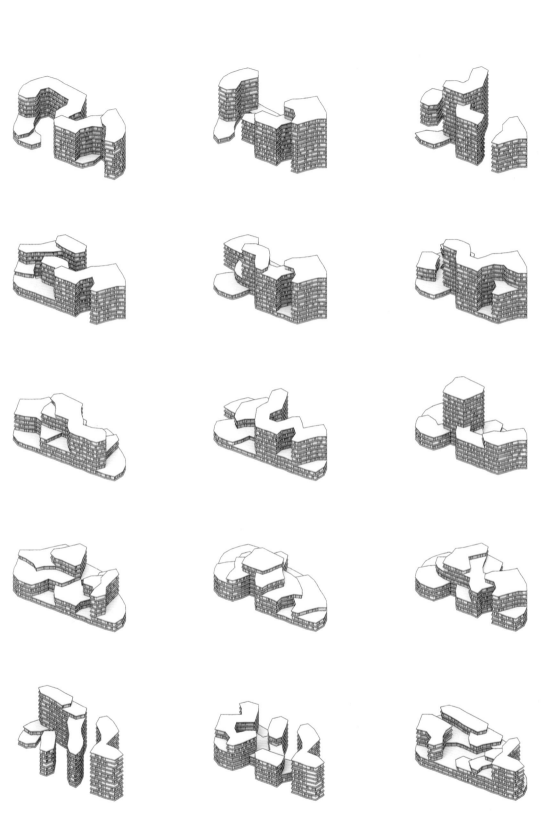

基于不同权重的生成结果（Different Weighted results）

重新生成的南京新街口地区 (Regenerated Xinjie kou District of Nanjing, China)

生成建筑渲染 (Rendering)

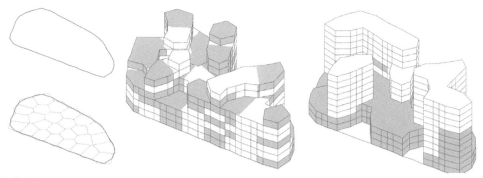

原始场地—细分场地—初始状态—优化后状态（Site-subdivision-mitialstatus-optimized result)

生成建筑渲染（Rendering）

泡与体 | Bubble and Volume

设计者：郭梓峰 | 指导老师：李飚
Designer: Guo Zifeng | Tutor: Li Biao

建筑布局设计是建筑设计的主要任务之一。随着生成设计技术的出现，该任务可以通过直接修改三维建筑空间而非绘制二维图纸来实现。因此，一种基于给定设计任务书的空间建筑布局设计的生成方法应运而生。任务书中指定房间数量、功能和拓扑关系等详细信息，结果将在几分钟内生成。

该方法结合了多智能体拓扑关系生成和进化算法空间优化，多智能体系统首先生成符合拓扑关系的布局。多智能体系统中引入特殊的胶囊形智能体，它可以在交互过程中自动调整其长度和方向，用于表示线性空间，如走廊和楼梯。

由多智能体系统生成的结果被转换为网格系统以进一步优化。优化采用进化策略：布局将改变并被评估。糟糕的修改将被丢弃，更好的修改将取代原有的结果。优化进程将不断进行，直到手动停止或满足所有预设要求。

Building layout design is regarded as one of the major tasks in architectural design. With the emergence of generative design technology, it may be accomplished by direct modification of architectural space three-dimensionally instead of making two-dimensional drawings. Thus, a generative method for spatial layout based on given programs is proposed. Detailed information such as room number, function and topology are specified in the program and results will be provided within minutes.

The proposed approach combines a multi-agent topology generative system with an evolutionary algorithm of spatial optimization. The multi-agent system first generates topology-satisfied layouts and the results will be further optimized. A special capsule-shaped agent is introduced in the multi-agent system, which adjusts its length and orientation automatically during the interaction. It is representing linear spaces like corridors and staircases.

The results generated by the multi-agent system are converted to a grid system for further optimization. The optimization adopts an evolutionary strategy: layouts will be changed and evaluated. Worse results will be discarded and better one will replace the old. The optimization continues until being manually stopped or all pre-set requirements are satisfied.

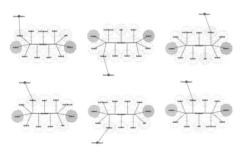

相同拓扑关系生成的不同结果（Multiple results based on the same topo logy)

多智能体系统：拓扑关系是空间布局的基本要求，如果房间之间的连接丢失，则导致布局无效。涉及多代理拓扑优化的目的是减少发生错误拓扑关系的可能性。系统将房间视为与泡泡类似的智能体。各智能体通过彼此交互调整其位置。交互规则缩短相邻物体之间的距离，同时防止彼此过于接近，并避免流线交叉。

进化优化：由多智能体系统生成的布局转换为网格模型，并进一步优化。转换采用Voronoi图的原理。优化采用模拟退火算法作为主要方法，即较差的修改也有机会被接受。评估过程由几个评价函数构成。

Multi-Agent System: Topology is the fundamental requirement for layout design, if the connections between rooms are missing, the layout becomes ineffective. The purpose of involving the multi-agent topology optimization is to reduce the possibility of incorrect topology. The system regards rooms as bubble-liked agents. Agents change positions by interacting with each other. The interaction rule is to shorten the distance between adjacent agents while preventing them from being too close to each other and avoiding intersections of circulation.

Evolutionary Optimization: The layout generated by the multi-agent system is converted into grid model for further optimization. The conversion adopts the principle of Voronoi Diagram. The optimization takes the Simulated Annealing Algorithm as its major approach, which means the worse revision may also be accepted. The evaluation process consists of several evaluators.

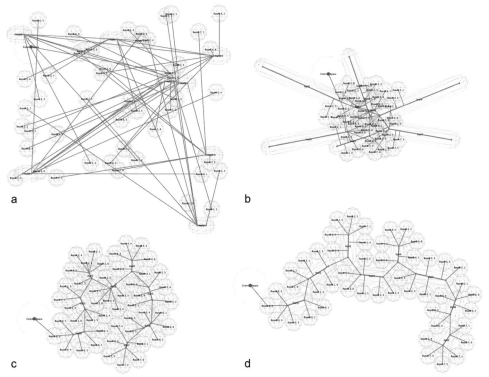

多智能体系统生成实例，由初始状态（a）到稳定状态（d）[An example of the multi-agant system, from initial stayus(a) to stable status(d)]

包含 144 个房间的生成结果（Result that contaits 144 rooms）

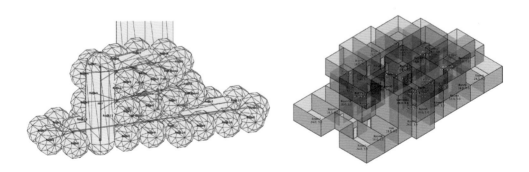

3 层住宅对应的 5 组结果（5 results for a 3-level house）　　2 层住宅对应的 12 组结果（12 results for a 2-level house）

生成过程：拓扑关系生成至体量优化（General process: topology, volumn and model）

10

基于 RhinoScript 的视线控分析及其应用
——教敷营地块建筑设计视线分析为例

The Sight Line analysis base on Rhino Script and Its Application
——Case study on the sight line analysis of Jiaofuying block

设计者：李力，UAL 工作室
Designer: Li Li, UAL Studio

教敷营地块将建设一个新的商业街区，因该地块紧邻著名的古典园林——瞻园，故在最大化建筑体量的同时，需保证园内的景色不受影响。鉴于二维图形视线分析不足以充分利用空间，我们开发了一个三维的视线分析工具来计算建筑后方的视线遮挡区域，又称"体积阴影"。我们在犀牛脚本中实现了该算法，并将其拓展为能同时计算多个视点。所有视点体积阴影的交集便是教敷营地块中对瞻园景观没有影响的建筑体量。

There is a new commercial district planned in the Jiaofuying block next to a famous classic garden, Zhan Yuan. We have to maximize the building volume without disturbing views inside the garden. The 2D sight line analysis commonly used is not precise enough to make full use of the space in this case, so a new 3D sight line analysis tool is developed to calculate the volume that blocked by the building from certain viewpoint, which is also called Shadow Volume. We implement this in Rhinoceros script and expand it to work on multiple viewpoints simultaneously. The intersection of all the shadow volumes will be the invisible area from Zhan Yuan which blocks no landscape views.

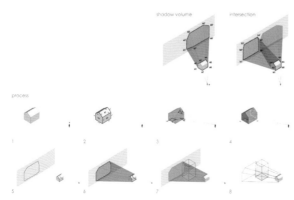

Algonthms

site plan and view points

bird view

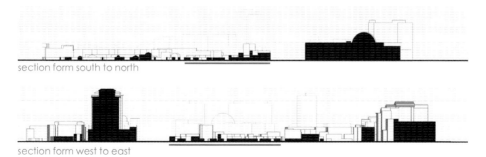
section form south to north

section form west to east

Site

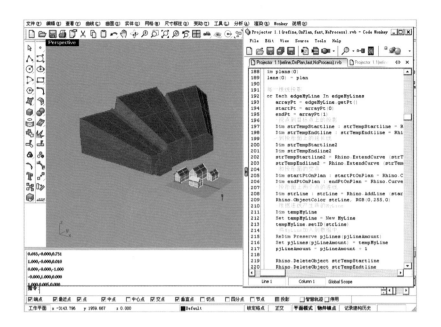

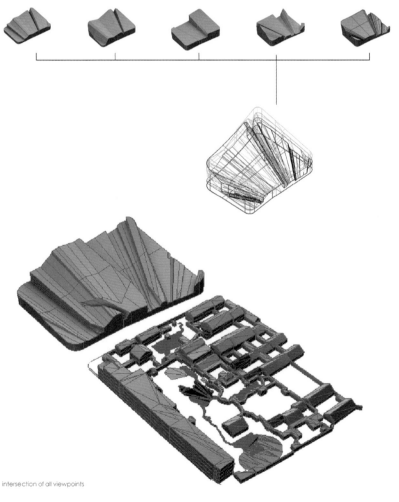

intersection of all viewpoints

Generation

切片模型

viewpoint A   viewpoint B   viewpoint C

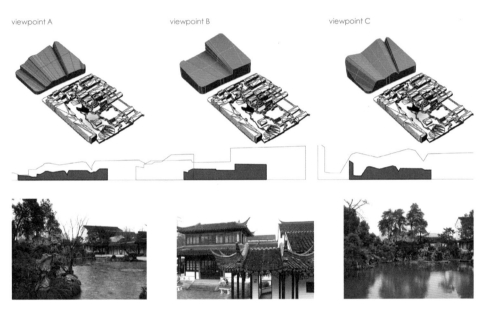

viewpoint D   viewpoint E

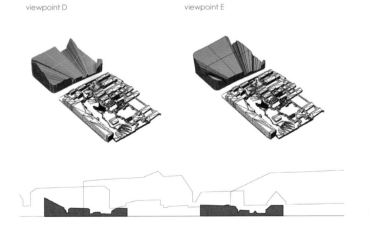

Shadow volumes

基于遗传算法的建筑体型优化
——以扬州南门遗址博物馆形体设计为例

The Optimization of Architectural Shape Based on Genetic Algorithm
——Case study on the design of Yangzhou South City-Gate Ruins museum

设计者：李力，UAL 工作室
Designer: Li Li, UAL Studio

遗传算法已被广泛地应用在优化领域，大量的研究都集中在如何提高算法的优化性能。但是面对实际建筑项目时，如何将设计问题转化为适合遗传算法处理的数据模型对最终的优化效果起到至关重要的作用。目前为止还没有一个同行的转化法则可以遵循。通常来说，遗传算法可以处理的问题可以分为数值问题和组合问题。在扬州南门遗址博物馆形体设计中，形体多边形的组合方式及次龙骨的分布方式对结构和构造的设计都有重要的影响，研究尝试将这个问题分解成组合和数值问题两个问题，并逐个解决。同时，探讨了遗传算法在解决此类问题的优势与劣势。

Genetic Algorithms (GA) is widely adopted in optimization and many researchers are focusing on how to improve its optimization performance. But when encountering real architectural projects, how to convert design problems into mathematical models that can be handled by GA is of great significance to achieve final optimization results. But so far, no such rule that can be used to guide this conversion has been developed for us to follow. In general, problems which can be handled by GA can be divided into combinatorial problems and numerical problems. In the design of Yangzhou South City-Gate Ruins museum, the structure and construction are strongly affected by the composition of polygon and the layout of sub frame. By trying to break down this complicated architectural problem into combinatorial and numerical problems, we are able to solve them one by one. Meanwhile, advantages and disadvantages of GA in this kind of problem are discussed.

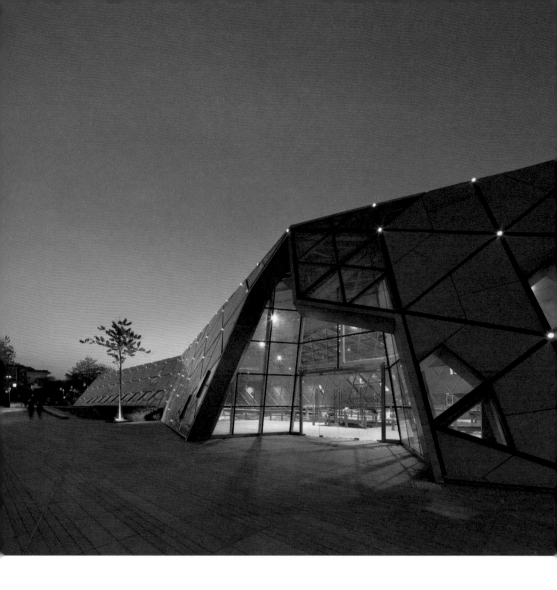

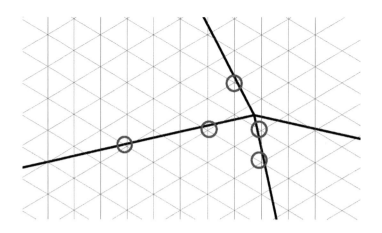

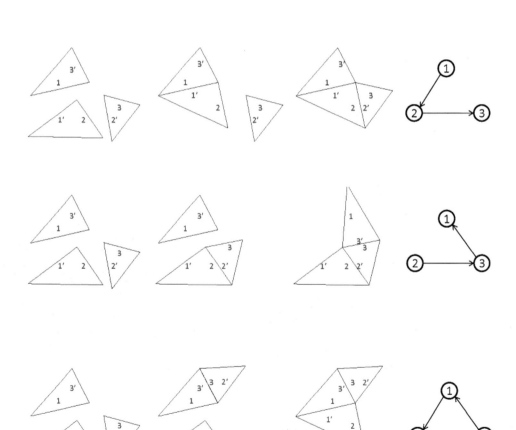

Algorithm

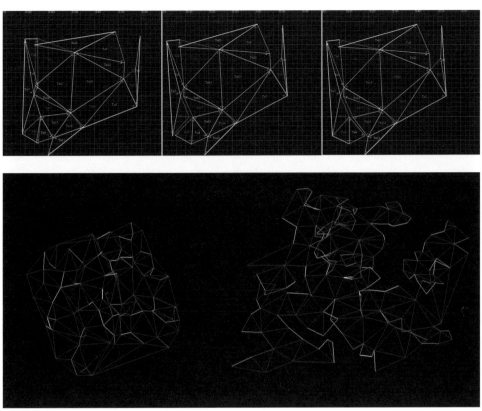
Unfolding algorithm

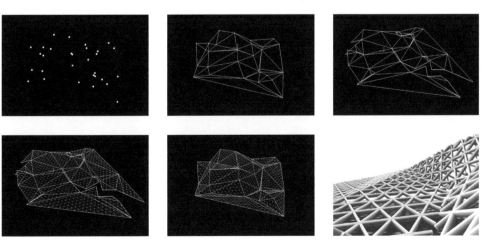
Optimigation

数控建造
Digital Fabrication

数控建造 (digital fabrication) 是一门与材料科学、制造工艺、建筑设计、结构力学、CNC (Computer Numeric Control) 技术密切相关的交叉科学。欧洲文艺复兴之后建筑的设计与建造逐渐分离，但是自 2000 年以来的数控建造技术历史性地把设计与建造融合起来，使建筑的数字化与物质化 (materialization) 获得了空前的统一。

数控建造通过计算机编程把设计概念转化为理性逻辑，并用数据驱动数控设备来完成加工与建造过程。由"设计—编程—制造"构成的数字链促使建筑师探索新的设计哲学与方法。运算化设计与建造 (computational design&construction) 为当今的建筑学提供了一个全新而系统化的视角来研究形式、材料、结构等建筑元素。数字化设计不仅可以提高设计的效率和建筑的品质，也能推动设计师发现新的设计问题并建立新的设计方法。

"建筑运算与应用研究所"每年开设"数控建造"本科设计课程，主要学习内容包括：Java 编程，计算几何 (computational geometry)，数控加工（CNC）以及数字建构 (digital tectonics) 方法。学生通过编程进行设计与建造，用程序逻辑来组织设计、加工、搭建等过程。课程强调材料与机器的行为特点（material performance & machine behavior）在设计中的积极作用。以 2016 年的数控建造课程为例，为期八周的课程安排如下：

第一周：学习 Processing 编程软件。任务：完成一幅生成艺术（generative art）作品。

第二周：了解激光切割技术；学习多边形、向量等几何图元的编程方法。任务：设计并制造一个平面构件，使用 Processing 输出 CAD 图形并用激光切割机进行加工。

第三周：学习计算几何；了解 CNC 泡沫切割机。任务：用程序生成泡沫切割机代码，切割一个三维物体。

第四周：学习悬链线（catenary）力学模型，了解构筑物的连接构造。任务：设计并制造一个立体构件，由平面构件经过折叠或拼接而成，尝试用同一类构造解决所有连接问题。

第五周~第八周：每个小组设计并制造一个构筑物。构筑物具有较大的内部空间，能够长期站立在室外。用程序生成构筑物的形态和细部，输出加工数据，用数控设备制造构筑物的构件并完成组装与搭建。

近十几年来，数控建造方法快速地在科研、教育、实践三个层面上并行发展。科研的主要任务是发掘数字化建造的潜能，把其他科技领域的优秀成果引入到建筑学中，并推动材料、自动化、计算机等相关领域的创新。教育的主要任务是建立系统化的理论与方法，引导学生们掌握并更新数字化设计与建造方法，培养具有跨学科视野的新型创新人才。实践的主要任务是用数字技术来推动制造业和建筑业的革新，并根据实践效果和市场反应对建筑数字化技术做出积极的调整。

Digital fabrication has the potential to unfold a new method of generative design. It creates an iterative design process that continuously integrates material, science, architectural design, structural engineering and manufacturing technology. Architects have allowed the separation between design and construction to go unquestioned since the Renaissance; however, the rise of digital fabrication technologies since 2000 has presented a historic reunion of architectural digitalization and materialization.

To formalize design ideas, we developed computational models that do not only encapsulate the design logics, but also incorporate the procedure for their digital fabrication. We investigated the complex characteristics of the material world and activated them as major stimuli to the design process. Beyond the preconceived rules of architectural design, now we can synthesize new thoughts in algorithmic logics and translate them into constructive processes with materilas. By coupling digitalization and materialization, computational design & construction brings a consistent network of novel methods and technologies for almost every aspect of architecture. The aim is not only to improve the quality and the efficiency of manufacturing, but also to open up new design problems and solutions.

The Institute of Architectural Algorithms & Applications has offered the undergraduate design studio of Digital Fabrication every year at Southeast University. The studio features Java programming, computational geometry, CNC fabrication and digital tectonics. The students employ programming tools to design and make prototypes. Programming logics become essential in organizing design, fabrication, and assembly. The studio is interested in how material behavior & machine behavior contribute to design innovations. The schedule of the 2016 studio is as follows:

Week 1: Learning Processing. Task: create a generative artwork through programming.

Week2: Learning laser cutting, and the programming of computational models of Euclidean vectors and polygons. Task: design and produce a planar component by a laser cutter. The cutting data must be created in Processing.

Week 3: Learning computational geometry and the 5-axis hot-wire cutter. Task: design and produce a piece of EPS foam artwork using the CNC cutter. The cutting data must be created in Processing.

Week 4: Learn catenary model through programming, and survey the connection details of constraction. Task: design and produce a 3D object out of multiple planar pieces by folding or nesting. Try using a parametric geometry for all connections.

Week 5-8: Each team (typically three students) design and make a pavilion that is spacious and stable enough in an outdoor environment. The students should use programming tools to generate form, connection details and output the data for CNC fabrication.

Digital fabrication has grown rapidly on the frontier of research, education and practice of architecture since 2000. The research explores the hidden potentials of digital fabrication and brings cutting-edge technologies from other disciplines to architecture. The interdisciplinary research also stimulates the developments in material science, automation, and informatics. The education establishes methods and theories of digital architecture and guides the students to understand and update the digital methods. We encourage the students to find their own interests within a wide spectrum of scientific fields around digital fabrication. The practice is to use digital technologies to renew the industry of built environment and to reshape the technologies according to the feedbacks from markets and costumers.

01

Angle-X

设计者：2009 年生成设计组（本科四年级：徐文，高浩，谢晓晔等）| 指导老师：李飚，华好
Designer: Digital Fabrication Team 2009 (Grade 4: Xu Wen, Gao Hao, Xie Xiaoye et al.)
Tutor: Li Biao, Hua Hao

Angle-X 探索了实际环境中的构筑物生成方法，并用数控建造的方式完成了构筑物的数控加工与现场搭建。课程从设计到施工历时八周。Angle-X 包括主结构及嵌板两部分，主结构由 13 个平面多边形构成。这些不规则多边形的空间位置充满了随机性，但采用了迭代优化的算法对其力学性能进行了优化。

方案设计从既有环境、人行流线及空间效果出发，进而制定了构筑物形态的优化规则并把这些规则编写成计算机程序（基于 Java 语言）。该程序能够优化构筑物的结构，确定所有连接构造的几何形状，并将加工数据传输给 CNC 激光切割机。激光切割机精确地加工所有构件，最后设计小组在现场把所有构件组装成 Angle-X 构筑物。Angle-X 是一个彻底的数字化工程，从设计到建造都无需图纸。

The Angle-X pavilion (2009) is one of the earliest digital fabrication works of Inst. AAA. The pavilion serves as an entrance for the exhibition hall of School of Architecture. The pavilion is made of carbon steel (5mm) as the main structure and stainless steel (1mm) as the skin.

Form-finding, structural engineering and CNC fabrication are highly integrated in a computational framework of this project. A hill climbing algorithm is employed to optimize the structure for more than ten criteria. The optimization process leads to a novel form with a sound structure, and subsequently generates the geometric data of all components including the structural elements, the connections, skin plates and all holes for screw bolts. These geometric data are directly transmitted from the computer program to the laser cutter for production. This workflow features a seamless digital chain from design to fabrication without any traditional drawings involved.

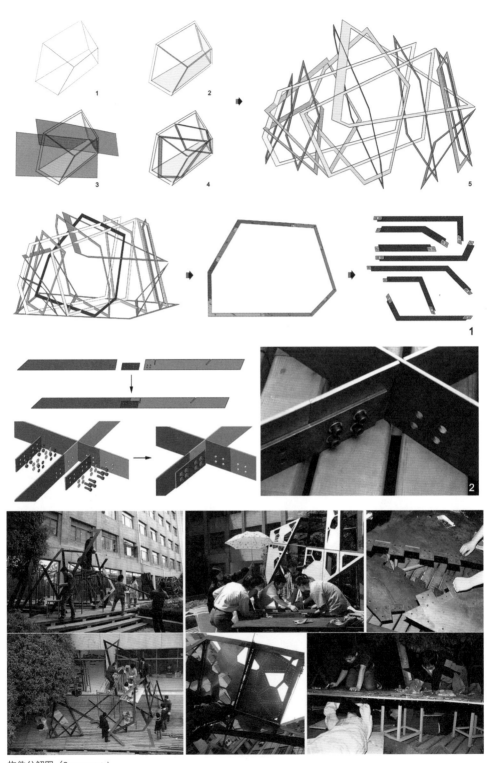

构件分解图（Components）

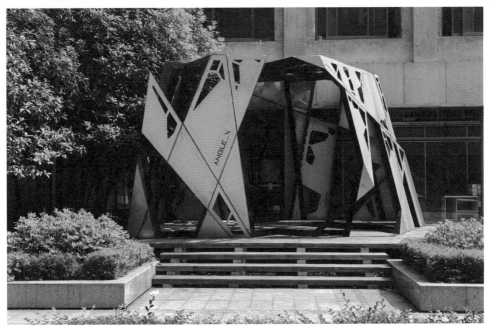

02

Cell0046

设计者：2011 年生成设计组（本科四年级：姚曜，宋心珮，何涛波等）| 指导老师：李飚
Designer: Digital Fabrication Team 2011 (Grade 4, Yao Yao, Song Xinpei, He Taobo et al.)
Tutor: Li Biao

Cell0046 项目研究了三维 Voronoi 算法，在三维空间中生成多个不规则"细胞"，整个构筑物由两个可分离的部分构成。每个细胞由平面不锈钢折叠和拼接而成。基于计算几何（computational geometry），该小组在自己编写的计算机程序中把每个三维细胞转化成平面构件。相邻细胞的开孔相互对应，以便于组装时进行精确的锚固。每个构件的平面形状数据由程序生成，数据驱动激光切割机进行精确的加工。最后设计小组把折叠形成的三维细胞拼装成完整的构筑物。

Cell0046 project employs a 3-dimensional Voronoi algorithm to create irregular cells within an arbitrary volume. The installation consists of two separable parts. Each cell of the installation is made of folded stainless steel. The 3D subdivision, the connection details and the data for fabrication are all created by the computer program developed by the team. Based on computational geometry, the team develops an algorithm to unfold each 3D cell into planar shapes. The openings and holes of adjacent cells are coupled in the computer program, which facilitates the precise assembly. The computer program generates the cutting data for each component and consequently the laser cutter produces all planar pieces. Finally, the team folded the cut pieces into 3D cells and assembled all cells as the final installation.

03

Tri·V

设计者：2013 年生成设计组（本科四年级：彭文哲，闫辰雨，许碧宇等）| 指导老师：李飚
Designer: Digital Fabrication Team 2013 (Grade 4, Peng Wenzhe, Yan Chenyu, Xu Biyu et al.)
Tutor: Li Biao

Tri·V 采用了悬链线（Catenary）力学模型、三角划分等算法，利用 Kuka 机械臂加工泡沫塑料构成穹顶状的构筑物。该项目探索了网状结构的力学优化过程：先用自组织（self-organization）方式生成顶点距离相等的三角网，再根据场地形状选择网状结构的锚固点，然后模拟悬链线的力学特性对网格进行形态优化，最终生成无剪力（compression-only）的曲面形态。该小组通过 Java 编程，把每个单元体的几何数据转化为 Kuka 机械臂的加工数据，由机械臂实现精确的热线切割加工。由于采用了数字化的设计与建造方法，最后的搭建过程高效而准确，用系统化的方式实现了复杂形体的数字化搭建。

Tri • V pavilion employs the catenary model and triangulation algorithms to generate a complex dome shape. A Kuka robot is used to cut all components for the final assembly. The project explores the optimization process of the mesh surface. First, a self-organizing process creates a triangular mesh whose nodes are evenly distributed within a given boundary; Second, the team chose the anchor points of the mesh surface based on the site configuration; Finally, an iterative process generates a compression-only surface with the predefined anchor points. All the generative processes are programming by the team with Java programming language. A hot-wire cutter is mounted to the robot flange so that the robotic system can cut the desired components out of EPS foam. The resultant components are glued together to form the pavilion. Thanks to the digital chain from design to fabrication, the assembly process is accurate and efficient.

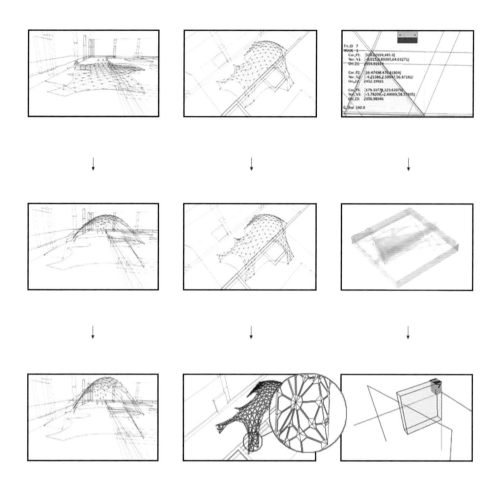

生成过程（Generative process）

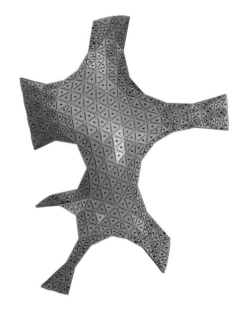

88

04

Dome · V

设计者：2015 年数控建造小组（本科四年级：孙世浩，祖丰楠，王衔哲）| 指导老师：李飚，华好
Designer: Digital Fabrication Team 2015（Grade 4：Sun Shihao, Zu Fengnan, Wang Xianzhe）
Tutor: Li Biao, Hua Hao

Dome·V 项目利用悬链线（Catenary）力学模型、Voronoi 等算法，将"数字链"系统方法应用于复杂形体的生成和数控加工。构筑物的造型、节点细部的生成和数控加工数据均通过 Java 编程来实现。类似穹顶的空间结构经过了力学优化，使表面只有压力而没有剪力，从而减少构件的形变、增强整体结构的稳定性。

穹顶表面利用 Voronoi 多边形进行剖分，每个多边形进而衍生出三维的单元体。计算机程序把三维单元体展开成平面形状，激光切割机把不锈钢板切割成相应的平面构件。这些平面构件经过弯折构成三维的单元体，最后将这些单元体拼接成穹顶状的构筑物。在这个数控建造项目中，学生们把形式逻辑完整地转化为数理逻辑，并设计了适合数控加工的节点构造，用数字化的方式完成了作品的设计、加工与搭建。

Dome·V applies the method of "digital chain" to form finding and CNC manufacturing. The optimization of the surface, the geometry of the connections and the data for production are integrated in a Java program. The pavilion combines the Catenary model and Voronoi tessellation, which facilitate structural optimization and surface subdivision, respectively.

The computer-generated solutions are performance-oriented: compression-only, pedestrian accommodation, appropriate sizes of cells and so forth. However, the resultant geometry is more or less unpredictable. A customized folding scheme transforms a thin panel (1mm stainless steel) into a stable 3-dimensional cell. The team develops an algorithm to unfold the desired 3-dimensional cells into 2D panels (as CAD files for laser cut). Through this project, the team learned to transform their design concept into computer logics and to further translate them into material, constructive processes.

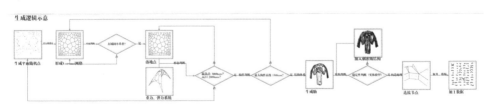

生成过程（Generative process）

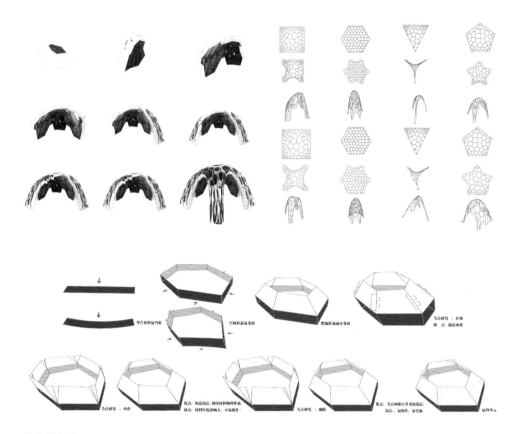

构造研究 (Study of tectonics)

95

05 Canopy

设计者：2015 年数控建造课程（本科四年级：徐静怡，俞敏浩，陈孜，郝子宏）| 指导老师：华好，李飚
Designer: Digital Fabrication Team 2015 (Grade 4: Xu Jingyi, Yu Minhao, Chen Zi, Hao Zihong)
Tutor: Hua Hao, Li Biao

Canopy 采用了从"单元"到"整体"的构建方式，选择钢板和薄膜相组合的构造方式。方案结合悬链线（Catenary）结构和细胞分裂的原理，以 Processing 为编程平台生成结构骨架，再以张拉膜为表皮，创造出形似华盖的遮蔽物。主结构采用 1mm 厚 201 不锈钢板，每根骨架用两片不锈钢片叠合而成，并向相反的方向弯折以增加强度。节点连接件为 2mm 厚 201 不锈钢，双层的扇形连接件可以固定每根骨架的位置和角度。外层表皮采用 PVC 张拉膜并用钢索固定。Canopy 项目采用了运算化的设计与建造方法，确保了从设计、加工到建造的系统性与精确性。

材　　料：1mm/2mm 201 不锈钢板，650g PVC 张拉膜
加工设备：激光切割机，折弯机
材料成本：6164.60 元
加工成本：4167.20 元
加工时间：15 小时
组装时间：36 小时

The Canopy project investigates the self-organization of the structural elements in a computational design process. The local logic of the Catenary model is integrated with a subdivision algorithm to create the pavilion's surface. The team employs the Java programming language to generate the pavilion's form and to write the geometric data for the CNC laser cutter. The pavilion is made of irregular beams of stainless steel that span the PVC membranes in between. The 201 stainless steel (1mm thick) is folded by a CNC bending machine to increase the stiffness of the beams. The method of computational design & fabrication seamlessly integrates the process of design, manufacture and assembly.

Material: 201 stainless steel (1mm/2mm thick), PVC membrane (650g)
CNC device: laser cutter, CNC bending machine.
Material costs: 6164.60 CNY
Manufacture costs: 4167.20 CNY
Manufacture time: 15 hours
Assembly time: 36 hours

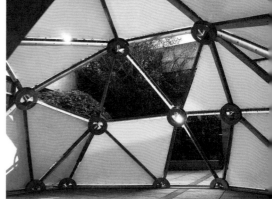

构件（Components）

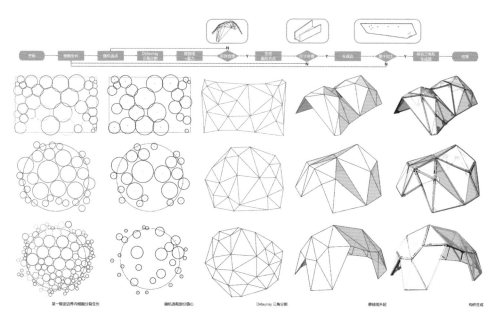

生成过程（Generative process）

06

Hakuna Matata

设计者：2016 年数控建造小组（本科四年级：张祺媛，刘苗苗，徐雨洁）| 指导老师：华好，李飚
Designer: Digital Fabrication Team 2016 (Grade 4: Zhang Qiyuan, Liu Miaomiao, Xu Yujie)
Tutor: Hua Hao, Li Biao

该项目利用弯折的铝材搭建了一个具有四个出入口的异形构筑物。整个方案从形体设计到加工数据的生成全部由 Processing 编程来实现，充分体现了从设计到制造的数据一体化。Hakuna Matata 构筑物采用了 6061 型铝板，利用大型激光切割机加工每个单元体的平面轮廓。每个单元体经过弯折之后强度得到显著加强，最终增强了整个构筑物的稳定性。因为传统机床不能对金属板材进行任意角度的弯折，本设计采用了"由定长得不定角"的构造方式。每个构造节点通过 3 种长度的杆件固定在两个单元体的不同位置，从而构成单元体间的不同夹角，最终实现构筑物的自由造型。

Hakuna Matata is an aluminum pavilion of four openings. The team employs the programming tool Processing to generate the pavilion's form, the connection details and all the corresponding data for CNC production. The processes of design and fabrication are integrated in a highly-customized computational framework. The pavilion is made of 6061 aluminum panels (2mm thick) which are cut by a CNC laser cutter. Each panel is folded to form a more rigid component. Screws of several lengths are used to connect a pair of panels with various geometric configurations. The continuous differentiation of the pavilion's surface is implemented through the parameterized displacements between the panel and the screws.

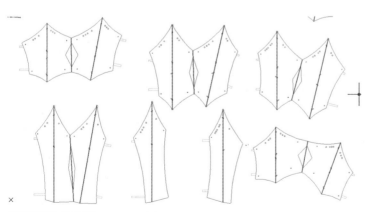

构件展开图（Unfolding components）

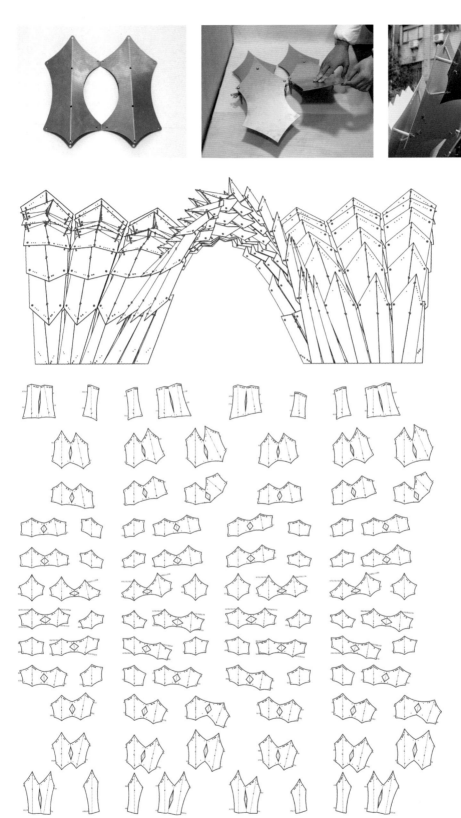

构件展开图（Unfolding components）

07

Neuron

设计者：2016 年数控建造小组（本科四年级：吴江源，梅琳丽，庞月婷）| 指导老师：华好，李飚
Designer: Digital Fabrication Team 2016 (Grade 4：Wu Jiangyuan, Mei Linli, Pang Yueting)
Tutor: Hua Hao, Li Biao

Neuron 项目利用编程方法生成纯压力结构（compression-only structure），兼顾美学与力学的需求，创造出一个形式复杂的薄壳结构。设计灵感来源于"neuron"（神经元细胞）：单元块（神经元细胞）之间通过各自突出的分支（突触）传递压力。类似神经细胞形态的单元块在减轻构筑物自重的同时创造了丰富的形态。

基于 Java 编程，该设计小组利用悬链线（Catenary）的自组织原理来生成整体形态。计算过程中保证每个质点的重力与单元块面积成正比，并且每个受力面都与其受力方向垂直。整个构筑物的表面只有压力，因此 Neuron 能够在没有粘结剂的情况下实现自支撑。单元体的几何形态和加工数据（G-code）由 Java 程序输出，实现了从设计到加工的完整数字链（digital chain）。单元体采用数控 5 轴热线切割机床进行加工，材料为高密度泡沫。切割过程中产生的泡沫负形可用来浇筑混凝土单元体。混凝土配方中加入了适量的珍珠岩以减轻重量。

链接：http://javakuka.com/plywood-shell/

The Neuron pavilion features a sophisticated compression-only structure generated by programming language Java. The design is inspired by the neurons which connect each other with synapses to bear pressure. The pavilion's elements adopt the shape of neuron to define a novel parametric geometry and simultaneously reduce materials and weight.

The surface's geometry is optimized through an iterative process to decrease the shear forces across the surface. Thus the entire structure is stable (in a static manner) even without any connections or glue between the components. A Java program is developed to generate the form based on catenary's self-organizing principle and create the G-code for a 5-axis hot-wire cutter. The CNC cutter produces all the neuron-shape components out of EPS foam. The components of concrete can be casted from the cut foam.

Link: http://javakuka.com/plywood-shell/

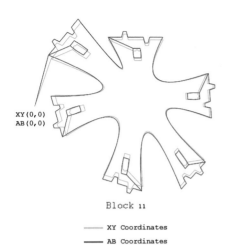

```
M03
M00
G01 X0 Y0 A0 B0 F400
G01 X611.2 Y235.3 A597.6 B430.9
G01 X616.4 Y221.4 A603.2 B404.9
G01 X621.3 Y208.3 A608.6 B380.4
G01 X625.9 Y195.9 A613.8 B357.5
G01 X630.3 Y184.3 A618.9 B336.3
G01 X634.5 Y173.4 A623.9 B316.6
......
G01 X502.5 Y147 A489 B342.6
G01 X548 Y184 A534.4 B379.6
G01 X570.1 Y168.3 A556.6 B363.9
G01 X581.7 Y177.8 A568.2 B373.4
G01 X571.2 Y202.9 A557.7 B398.5
G01 X611.2 Y235.3 A597.6 B430.9
G01 X0 Y0 A0 B0
M05
M30
```

XY(0,0)
AB(0,0)

Block 11

—— XY Coordinates
—— AB Coordinates

切割构件的 G-code（G-code of cutting component）

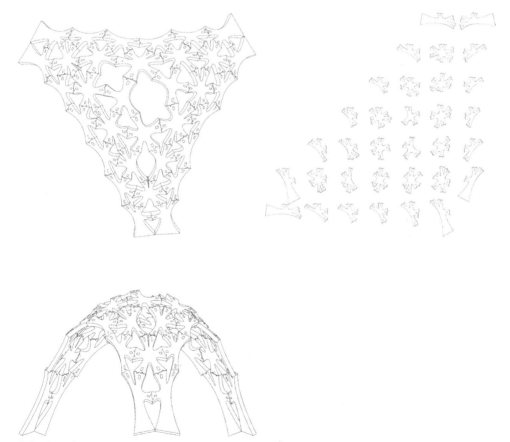

构件分解图（Subdividing the surface into neuron-shaped components）

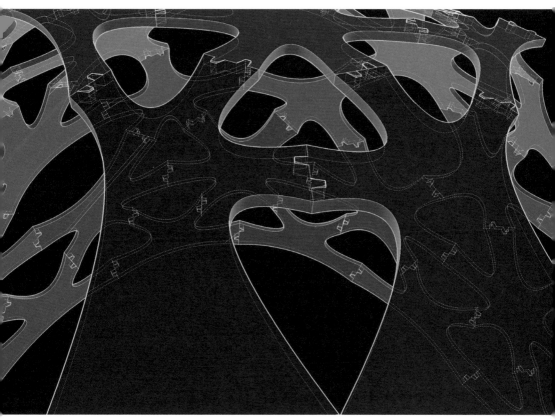

构造细部（Joints）

08

Visual Robot

设计者：2016 年数控建造小组（本科四年级：王嘉诚，陈嘉豪，吴晓涵）| 指导老师：华好，李飚
Designer: Digital Fabrication Team 2016（Grade 4： Wang Jiacheng, Chen Jiahao, Wu Xiaohan）
Tutor: Hua Hao, Li Biao

工业六轴机器人与传感器结合可以构成实时互动系统。Visual Robot 项目把 cognex 摄像头、机器人以及自动化机械爪三个部分结合起来。摄像头捕捉图像向机器人传递物体位置的信号，机器人到达相应位置并向机械爪传递脉冲信号，机械爪随即完成物体的抓取动作。参与者一旦移动物体（苹果），机器人系统将重新识别它的位置并进行抓取。这种机器人系统可以服务于互动式的数字化建造。

链接：http://javakuka.com/interact/

Connecting a camera to a serial manipulator enables the robotic system to observe its environment and react responsively. The low-level signal captured by the camera, namely the pixels, indicates the current position of the object (an apple in this case) and constantly informs the robot's motion. Once a user moves the apple, the robot will recognize the new position and grasp the apple. The team engineers the camera-robot communication, makes the gripper and its controller and programs the robot's motion. Such robotic system can facilitate the digital fabrication in a responsive manner.

Link: http://javakuka.cpm/interact/

KATCHER

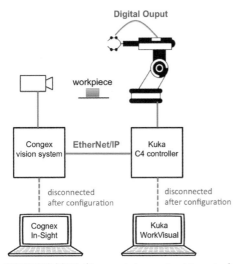

视觉系统组成部分（Framework of responsive vision system）

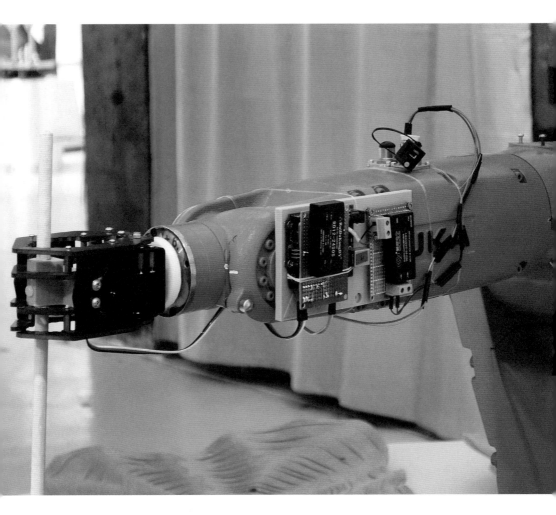

110

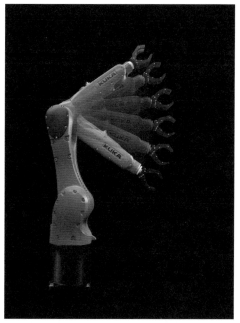

09

融·合（2014）| Harmony·Peace（2014）

设计者：李飚，郭梓峰
Designer: Li Biao, Guo Zifeng

"融·合"为南京理工大学60周年校庆所建。其造型借鉴了多个波源产生的波形相互干涉产生复杂形态的物理现象。该项目通过Java程序模拟波形的叠加，生成由水平和垂直构件插接而成的墙面，并创建所有构件的加工数据。数字化的方法贯穿了从设计到建造的每个环节。项目采用了激光切割机加工钢板从而获得所有不规则构件。通过改变波源的数量、位置及强度，便可得到形式各异的作品。参数化的构造设计兼顾了加工与组装的便捷性。

The "Harmony·Peace" installation is built for the 60th anniversary of Nanjing University of Science & Technology. It adopts the principle of wave propagation. The interference produces very complex waveforms when multiple sources are involved. The computer program simulates the interference of waves, generates the walls of horizontal and vertical strips and creates the data for fabrication. The computational method covers every stage from the design to the final construction. A laser cutter is employed to cut the complex strips out of metal sheets. Modifying the positions and the amplitudes of the wave sources will lead to distinct designs. The parametric geometry of connections between the horizontal/vertical strips is concerned with the convenience of assembly and the laser cutter's capability.

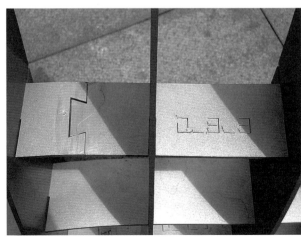

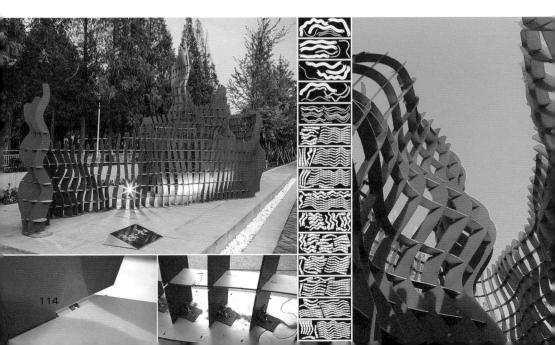

10

印象太湖石（2016）
Taihu Stone Imagination (2016)

设计者：郭梓峰 | 指导老师：唐芃，李飚
Designer: Guo Zifeng | Tutor: Tang Peng, Li Biao

"印象太湖石"位于宜兴太湖绿道，用运算化设计的方法抽象地再现了太湖石"瘦、透、漏、皱"的特征。印象太湖石基于Gyroid极小曲面，以元球（Metaball）为辅助塑形手段，生成充满孔洞、造型奇特的太湖石形象。在程序中调整元球的数量、位置及半径，便可实现造型的实时修改。

通过多次实验，"印象太湖石"最终采用了水平板片堆叠的建造方式。材料为5mm厚碳钢板，由激光切割机进行加工。所有螺栓的位置、构件的切分位置及构件排版均由程序自动完成。印象太湖石的实际尺寸为7.5m×2.8m×3.2m，由177块构件组成。激光切割总长度为802m，螺栓1060组。工厂加工时间为三天，现场组装时间为两天。

The "Taihu Stone Imagination" is built on the side of Taihu Lake at Yixing. The team developed a computer program to model the complex geometry of Taihu stones, which has been widely recognized as an essential symbol of Chinese traditional aesthetics. This design combines the Gyroid minimal surface and the meatball algorithm to create a complex geometry full of irregular holes similar to that of Taihu stones. Modifying the positions and the radii of the metaballs in the program will immediately lead to new results.

The installation is built by layering steel panels. All the steel components are cut by laser cutter. The geometry of each layer, the subdivision of components and the shape of connections are all generated by the computer program. The dimension of the installation is 7.5m×2.8m×3.2m. It consists of 177 steel components and 1060 bolts. The head run of the laser cutter is 802m. Laser cutting takes three days, while the in situ assembly takes two days.

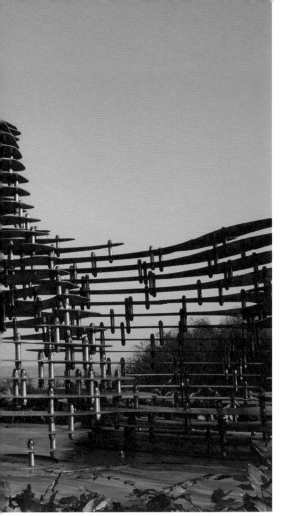

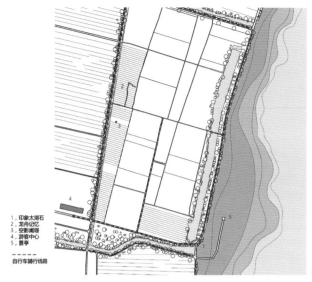

基地平面图（Site plan）

1，印象太湖石
2，龙舟记忆
3，光影阑珊
4，游客中心
5，景亭
----- 自行车骑行线路

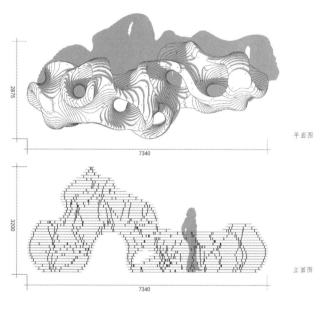

平面图

立面图

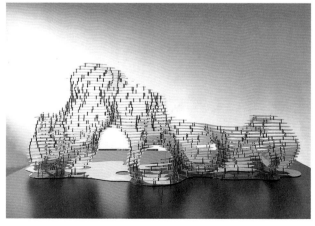

1∶5模型（1∶5 model）

空影阑珊 | Sparse Shadow

设计者：季云竹 | 指导老师：唐芃，李飚
Designer: Ji Yunzhu | Tutor: Tang Peng, Li Biao

"空影阑珊"景观小品位于宜兴太湖绿道。该设计结合了人在场地内的行走路线，让行人在运动过程中感受到"空影阑珊"呈现出来的多种视觉效果。作品主要由竖杆和挂片构成。竖杆的位置呈放射状空间排布，充分考虑了人眼的透视原理。竖杆上挂片的形状是由多个图像在空间中重叠而成。计算机程序首先运用 Blob Detection 算法分别处理多个图片，然后将多个平面图样转换到同一个三维空间中，再用像素化的方式对挂片进行开洞处理，使视觉效果更为精确和生动。

"Sparse Shadow" is located on the Greenway of Taihu Lake at Yixing. The pedestrians will observe the distinct aspects of the artwork when walking around the site. Several images are transformed and combined into the 3D volume of the installation. "Sparse Shadow" consists of 17 columns and 498 metal pieces that deliver a dynamical visual effect. The computer program first transforms the original images into bitmap-like layouts; then several layouts are superposed within the installation's volume and the visual effect is optimized; finally, the computer program creates the data for laser cutting.

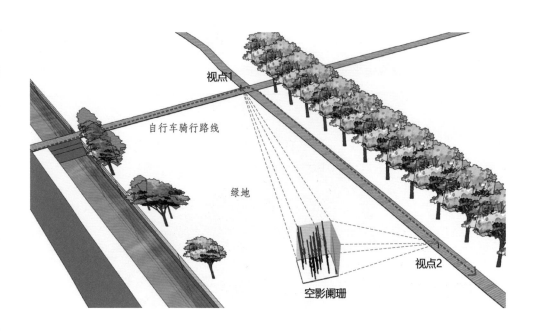

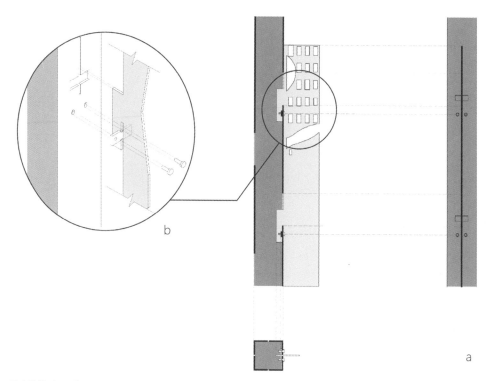

节点构造（Detail）

12

龙舟记忆
Dragon Boat Memory

设计者：张佳石 | 指导老师：唐芃，李飚
Designer: Zhang Jiashi | Tutor: Tang Peng, Li Biao

"龙舟记忆"在原有景观道路上创造出太湖泛舟的形象，整体造型与周围的花海融为一体。项目根据场地的现状，在基地上设定控制点后由NURBS(Non-uniform rational basis spline) 曲面生成整体形态。从形态设计到所有构件的数控加工，都由团队编写的Java程序来完成。构筑物由主体框架和穿孔板组成。主体框架由金属切片双向插接互锁而成，材料为5mm碳钢板，表面镀锌喷漆。附在主体框架上的穿孔板通过点阵拟合，将水墨画图像映射至三维曲面，既有防护作用，又能展现太湖及江南水乡的意境。穿孔板采用了1.5mm的铝板。主体框架和穿孔板都由激光切割机加工而成。

The "Dragon Boat Memory" installation creates a unique landscape on the side of Taihu Lake at Yixing. The shape of the installation is modeled by a NURBS (Non-uniform rational basis spline) surface whose control points are located on the site. The team developed a Java program to create the geometry of the installation and produce all the data for digital fabrication. The installation consists of a structural framework and punched panels attached to the framework. The framework comprises of horizontal and vertical beams which are interlocked by each other. The beams are made of galvanized steel (5mm thick). The punched panels create an abstract image of local traditional paintings. These panels also enclose an amusing passage for the visitors. The pands are made of 1.5mm aluminum sheets. Both the framework components and the punched panels are produced by a laser cutter.

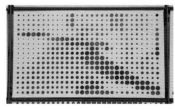 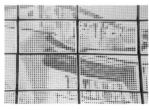 

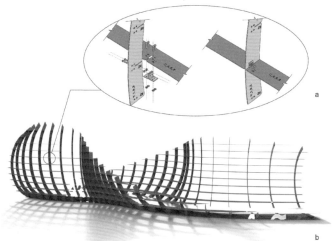

b 构造（Connections）

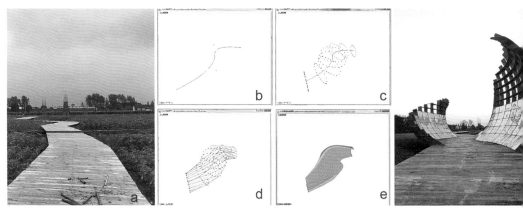

生成过程（Generative process）

13

Ceiling Margin

设计者：2007年生成设计组（本科四年级：郁倩，胡宏等）| 指导教师：李飚
Designer: Generative Design Team 2007（Grade4：Yu Qian, Hu Hong, et al.）| Tutor: Li Biao

Ceiling Margin 为东南大学建筑学院门厅吊顶工程，2007年建成，是建筑学院早期的数控建造实践。该实践项目运用编程方法完成形态设计与数控加工。Ceiling Margin 将吊顶设置为连续三维曲面，并合理避让现有管线，实现门厅的空间利用最大化。设计小组编写的计算机程序一方面能够生成吊顶的几何形状，另一方面能够产生所有构件的加工数据用于数控激光切割。从设计、加工到搭建，形成了一条完整的数字化生产链。

Ceiling Margin focuses on the form generation and production of a surface which covers the ceiling of the entrance hall at the School of Architecture, Southeast University. The continuous 3-dimensional surface keeps away from the existing pipes and beams on the ceiling. The team developed a computer program to generate the geometry of the free-form surface and create the data for laser cutting. The laser cutter produces steel and wood boards to produce all the components of the ceiling. This project constructs a complete 'digital chain' from design to production.

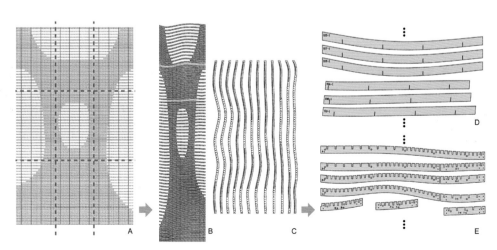

构造细部 (Details)

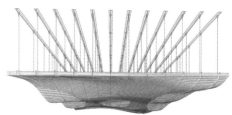

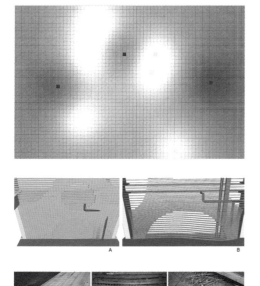

14

槃 | Panzi

设计者：华好
Designer: Hua Hao

　　工业机器人的非标准化应用是当代数控建造技术的主要特点。六轴机械臂的主要优势是加工方式的灵活性：在机械臂法兰盘上安装不同的末端工具（end effector）就能实现不同的加工工艺，如铣削、三维打印、组装等。

　　槃同"盘"，古字形暗示它由木材制成。基于现代人造板材和新型节点的开发，该项目探索了木制盘的数控加工工艺。配置了电动主轴的机械臂是进行铣削加工的主要工具。材料为中密度纤维板（Medium Density Fiberboard）。Ginkgo 节点将平面板材组合在一起构成一个三维物品。Ginkgo 节点呈连续的 S 形，由数学表达式来定义。该节点设计充分利用了密度板的轻微弹性，使榫卯处保持较大的摩擦力从而保证整体结构的稳定性。为了在最底层控制机械臂的运动路径，我们编写了 Java 程序来直接生成 KRL 代码（Kuka 机械臂专用代码），并发布了开源代码库 javakuka.com。

The Panzi project produces trays by cutting Medium Density Fiberboard (MDF) and connecting the cut pieces by Ginkgo joints. A robotic system is engineered for cutting and milling. In the Ginkgo joints, the parametric curve across the horizontal surface prevents the two connected plates from moving horizontally; while the tiny gears through the depth of the plates prevent vertical movements.

A spindle is mounted to the robot's flange so that the robot becomes a multi-axis milling machine. The Ginkgo joint is not orthogonal to the fiberboard's surface. That's why a robot instead of a conventional milling machine is employed. The project uses Java to program the tool path and write the geometric data into a KRL format (for Kuka robot). This project also publishes an open-source library (javakuka.com) for convenient programming of Kuka KRL in Java. There is no third-party software involved in this project.

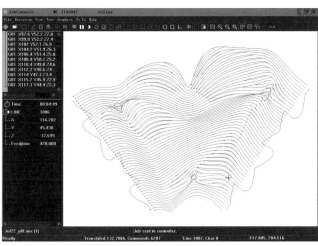

加工构件的 G-code
(G-code for the milling machine)

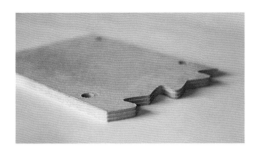

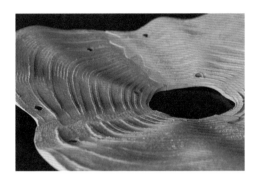
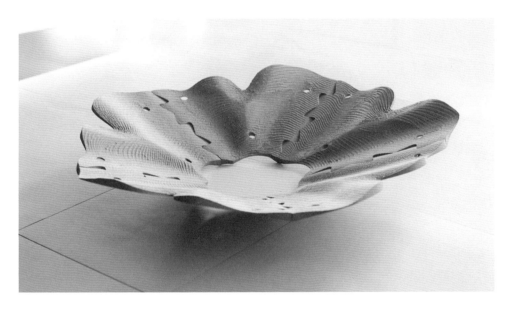

物理计算
Physical Computing

信息技术的发展已经影响了人们的生活及生产方式，尤其是传感器、微电子以及无线通信技术的进步，使得人与计算机的交互不再局限于桌面电脑。越来越多的数字设备以移动和嵌入的方式融入日常活动和环境中，催生了更丰富实用的交互应用方式，物理计算的概念也应运而生。虽然建筑系统因其复杂性与综合性，对新技术的响应会有一定的滞后性，但相关的概念和技术也开始逐渐渗透到建筑教育、建筑设计及建成环境之中，成为数字建筑研究的一个分支，国内外学术领域也已有一些尝试性的探索。

物理计算指通过软硬件开发来加强计算机与现实世界交互关系，从而促进两者的融合，是一个宽泛的概念。这里的计算机不仅指大型服务器、个人电脑、智能移动设备，同时也包括嵌入式处理器、芯片等一切具有运算能力的设备。这个概念随着近年来传感器、人机互动、无线网络技术的发展，以及软硬件开源开发平台的增加应运而生，不论在消费电子产品，或是工业化的生产流线，都有不同形式的应用。物理计算与信息、自动控制等专业领域常用的普适计算、嵌入式计算、环境智能等概念十分接近。相对而言，物理计算在非计算机专业人士中使用更为广泛，例如设计师、媒体艺术家、电子技术爱好者等。相对于技术研发，他们更关注创新性的运用、美学及社会意义。

物理计算的产生与发展主要依赖以下几个因素：硬件成本的降低，数据需求增加，以及学习曲线的缩短。硬件成本的降低，降低了研发和生产成本；大数据的需求，增加了对传感器等基于物理计算技术设备的需求；学习曲线的缩短，降低了学习成本，吸引了大批包括建筑师在内的非计算机专业人员的研究兴趣。

物理计算作为数字建筑的一个新的分支也正逐步发展。在欧美国家院校的一些建筑、设计、媒体专业，早在十多年前就陆续设立了相关研究机构和课程。东南大学建筑学院早在 2009 年就由李飚教授从苏黎世联邦理工学院 CAAD 教研组引入互动设计课程。

物理计算和建筑学的结合方式多样。首先，根据作用对象的不同，可分为交互式建筑设计过程和互动建筑两大类的研究。交互式设计过程作用于建筑设计方法本身，而互动建筑着眼于如何将物理计算技术应用于建成环境之中。根据使用技术和设计侧重点的不同，互动建筑又可分为以下几类：互动装置、互联建筑、智能建筑。

三类互动建筑都建立在互动设计的闭环系统之上，但各自的侧重点和技术难度有所不同。互动装置主要探索的是建筑中互动元素与人的互动形式，营造不同的视觉感受和空间体验。目前，这是各大设计院校互动课程的主要教学内容。互联建筑是指将建筑内的电子电器设备、传感器相互连接实现数据的交互，从而增强控制。随着无线网络技术的成熟，相关产品已经开始进入消费市场。智能建筑指运用大数据技术、人工智能技术提高建筑智能化程度，提供更人性化的服务。此部分内容还在研究与实验阶段。

　计算机和建筑相互融合的技术难度正在逐步降低。从建筑设计的角度出发，这些数字电气设备不应该再被视作建筑的附属物，而应当做是一种新的数字建材。在信息和智能化时代，建筑不仅提供空间场所，而应是一个与人互动的居住服务提供者。

The advent of information technology has changed our daily life. New achievements in sensor, microelectronics and wireless communication technology have led the human-computer interaction to a post-desktop era. More and more mobile and embedded digital devices are coming into our everyday life and environment, which enrich the interactions between human and computer. The concept of Physical Computing is born in such background. Due to the complexity and integrity of the building system, there is always a time lag responding to new technologies. However, Physical Computing begins to infiltrate architectural education, design and the built environment, and becomes a new branch of digital architecture. There are already some tentative research explorations in academic field both home and abroad.

Physical Computing refers to enhance the interaction between human and computer by developing software and hardware. Here, the computer is not only a server, PC or mobile device, but all computational devices that have processors and chips embedded. The notion of physical computing is formed with the developments in sensors, human-computer interaction, wireless communication technologies and the increasing of open source software and hardware platforms. Its applications vary from consumer electronics to industrial production lines. This concept is similar to Ubiquitous Computing, Pervasive Computing, Embedded Computing and Ambient Intelligence that are used in Informatic and Automation research field. However, physical computing is more widely used in non-computer professionals, such as designers and media artists. Compared to technology development, they concern more about its innovative applications, aesthetic value and sociological significance.

The development of physical computing relies on these three factors: the drop of hardware cost, the increasing demand of data and the shortening of learning curve. The lower hardware cost brings down the research and production cost; the demand of big data increases the demand of sensing devices. The shorter learning curve attracts research interest for architects and other non-computer professionals.

Physical computing, as a new branch of digital architecture, is still evolving to maturity. Some architecture, media, and design schools have established relative research institutes and courses in last ten years. Early in 2009, Prof. Li Biao introduced interactive design courses to the school of architecture of SEU from the CAAD of ETH.

There are plenty of different ways that physical computing technology can be integrated into architecture. Firstly, depending on the research target, it can be divided into research of Interactive Architectural Design and Interactive Architecture. Interactive Architectural Design focuses on the design method, while Interactive Architecture focuses on the building environment. Interactive Architecture can be divided into three subfields: Interactive Installation, Connected Building and Smart Building.

The emphasis and technical difficulties of these three subfields are slightly different. The Interactive Installation is mainly about discovering new forms of human-computer interaction, and novel visual and spatial experiences. Currently, it is the main teaching content of interactive courses in major institutions. Connected Building refers to connect all electronic devices and sensors together, enabling the data exchange and enhancing the control. With the development of wireless communication technology, some related products are entering the consumer market. The Smart Building refers to improve the intelligence of the building control system by introducing data mining and artificial intelligence technologies, which is still in research and experiment.

Technical difficulties of integrating computer technology with architecture are getting less. From the point of view of architectural design, these digital components should not be regarded as appendages of a building, but a new digitalized material. In this information and intelligence era, building not only provides space, but also services for better interactive living.

01

互动设计专题的最初尝试
Original Attempt of Interactive Design

设计者：2009年互动设计小组 | 指导老师：李飚
Designer: Interactive Design Team 2009 | Tutor: Li Biao

Arduino开源电子原型平台包含硬件Arduino控制板和软件开发平台（Arduino IDE），由欧洲开发团队于2005年开始研发。借助简单的编程逻辑，初学者可以在很短的时间内实现准商业级的互动案例。2009年，学院决定在大学四年级建筑设计课程的跨学科研究专题中加入基于Arduino套件及其外设的物理计算与互动设计课程，该课程同时也作为学生研习Processing(Java)程序的前导练习。师生在资金短缺的逆境中坚持探索，完成了"光盒"、"触屏"等四项互动案例。

Arduino is an open source electronic prototyping platform including hardware of Arduino control board and software development platform (Arduino IDE). It was researched and developed by a European team in 2005. With simple programming logic, beginners can achieve quasi-commercial interactive cases within a very short time. In 2009, our school decided to include a physical computing and interactive design course based on the Arduino suite and its sensors in an interdisciplinary research topic for the fourth year architectural design course, which would also serve as a prerequisite for students to study Processing (Java) programs. Teachers and students explored and completed four projects such as "light box" and "touch screen" in adversity of funding shortage.

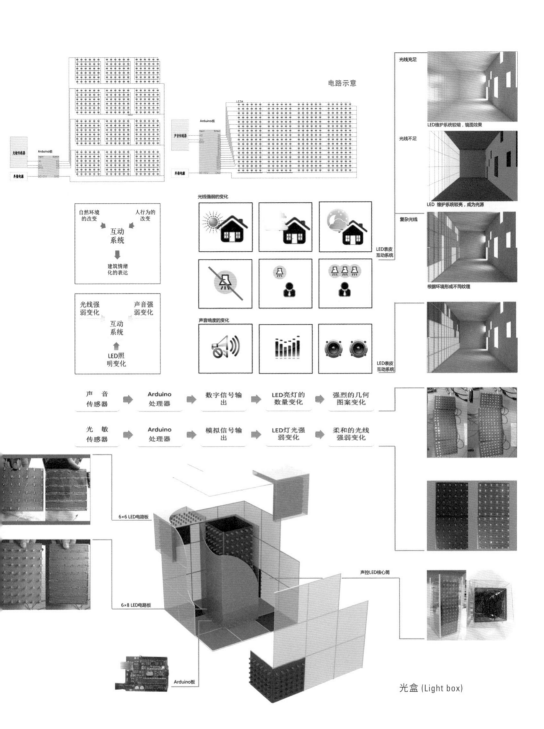

光盒 (Light box)

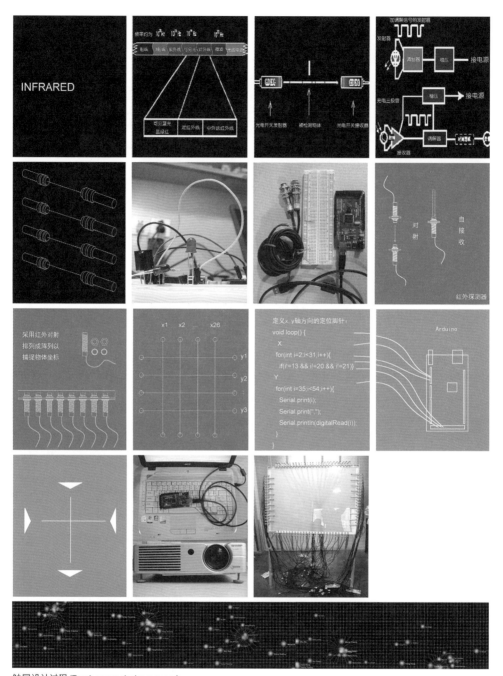

触屏设计过程 (Touch screen design process)

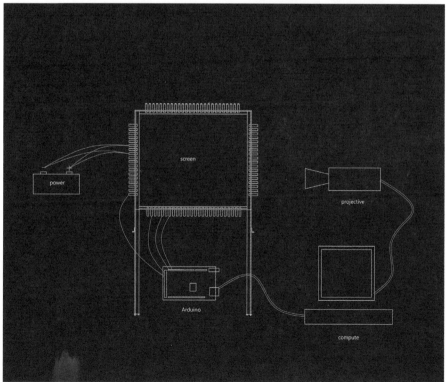

触屏 (Touch screen)

02

塑造景观
Constructing Landscape

设计者：李力 ｜ 完成时间：2012 年
Designer: Li Li ｜ Complete time: 2012

本研究尝试建立一个互动式的景观设计过程。设计者直接在小比例的场地沙盘上，以雕塑的方式来改变地形进行设计。修改过的地形经过扫描后输入电脑成为数字模型。研究开发了一个集成坡向、汇水、可视度等分析算法的软件，可实时对数字模型进行分析，并以可视化的方式反馈给设计者。设计者比较自己设计意图与分析结果的差别，进一步修改地形，如此形成一个循环往复的过程，设计者每一次修改后都能看到修改所产生的相应的影响。

This research attempts to establish an interactive landscape design process. The designer does his design by carving on the small-scale sand box. The modified terrain can be scanned into a digital model. A software integrating slope, water flow and visibility analysis has also been developed for real-time analysis and visual feedback. The designer can compare his intention to the analytical results and change the design accordingly. Such kind of feedback loop can help the designer see the consequences whenever a change is made.

汇水分析（Water flow analysis）

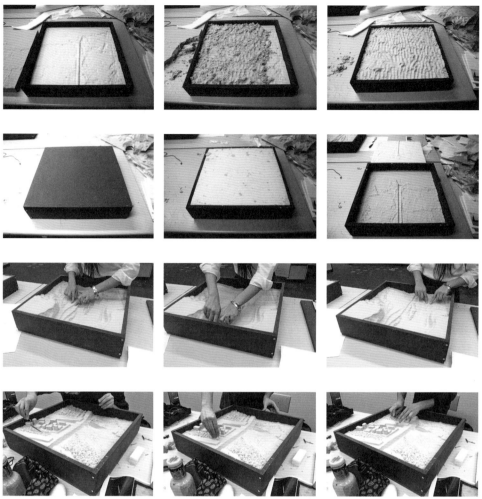

沙盒（Sand box）

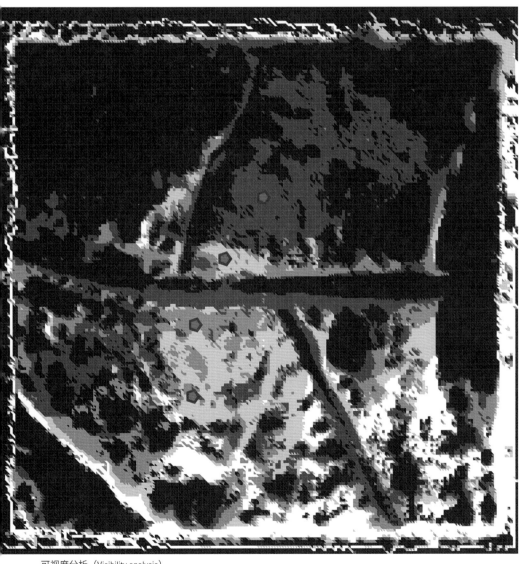

可视度分析 (Visibility analysis)

ND# 03

摆动的结构
Wiggling Structure

设计者：2014年互动设计小组：戴嘉熙，卓可凡 | 指导教师：虞刚，方立新
Designer: Interactive Design Team 2014: Dai Jiaxi, Zhuo Kefan | Tutor: Yu Gang, Fang Lixin

"摆动的结构"试图提出一套互动张拉整体结构系统，可以通过感应器接收输入信息，然后将处理过的信息传输到张拉整体结构，与人体行为产生互动。具体而言，通过超声波传感器根据人体行为将数据传输到控制系统，经过数据处理，再控制传动装置的收缩和舒张，实现结构的形变和运动，同时辅以相关设备调整变形幅度，使系统的整体运动幅度处于可控范围。最后，这个方案还设计了移动设备界面程序，可以让使用者通过远程控制模块调整结构形变何时开始、何时结束。这个模型原型可以作为相关的遮蔽结构原型，例如根据不同天气条件，为人们提供相应可变的互动遮蔽设施。

The Wiggling Structure tries to propose an interactive tensegrity structure, which is able to sense and react to human movement by changing its shape. In particular, it detects the approaching of people with the ultrasonic ranging sensor. After data processing, it controls the tension of the actuator to change the form of the structure. The adjustment of the changing is limited in a certain range to ensure all movements are under control. Additionally, an APP is developed for remote control. This model can be considered as a prototype for shading structure, for example, an interactive installation that can change shape according to weather condition.

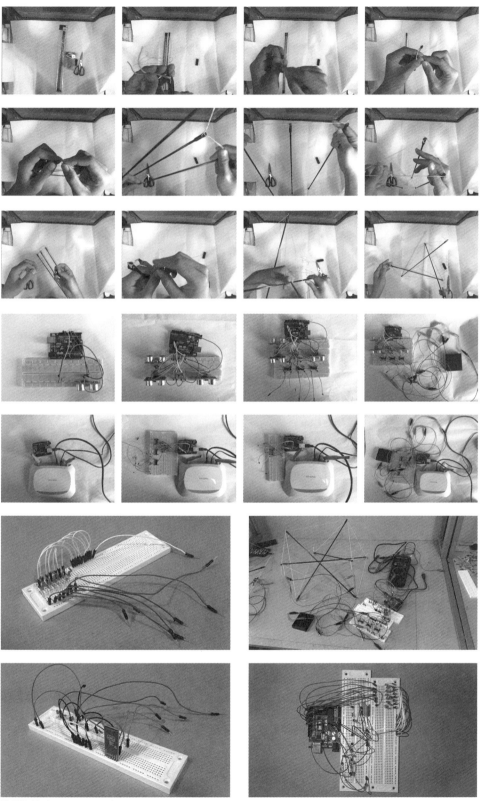

设计流程（Design process）

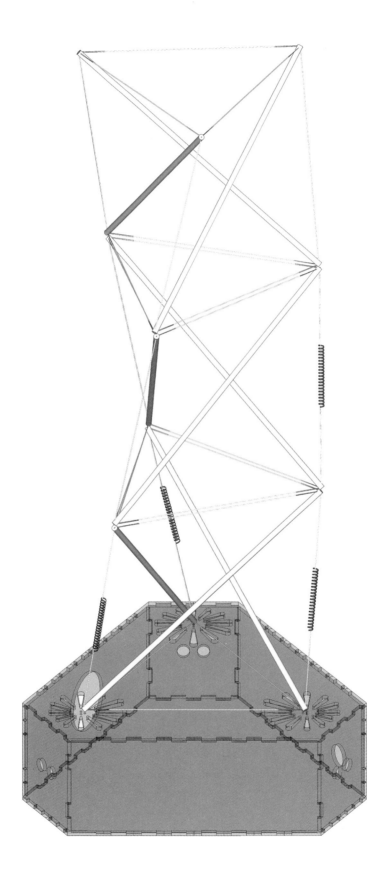

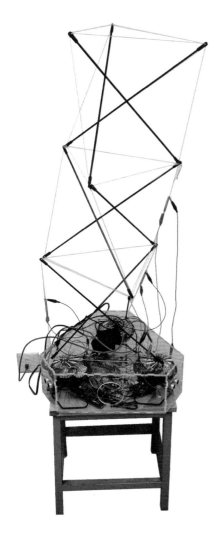

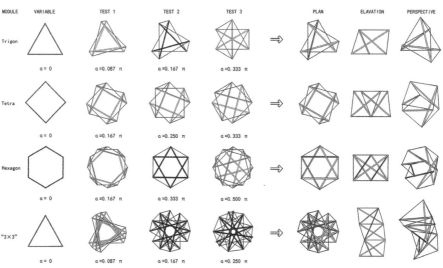

形态分析 (Form analysis)

155

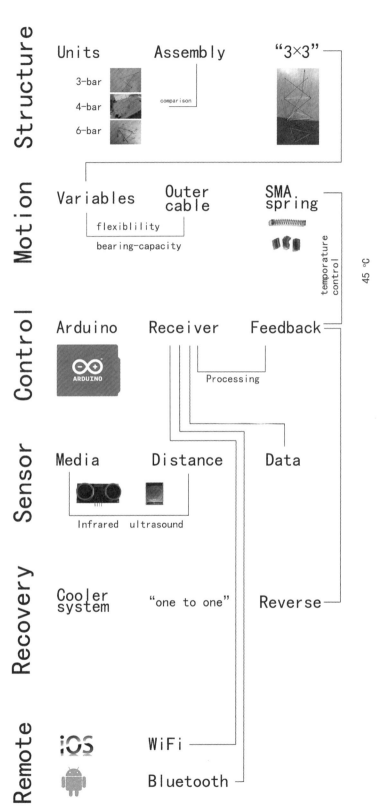

系统架构（System architecture）

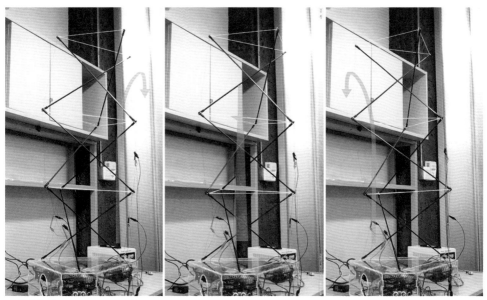

互动过程（Interaction process）

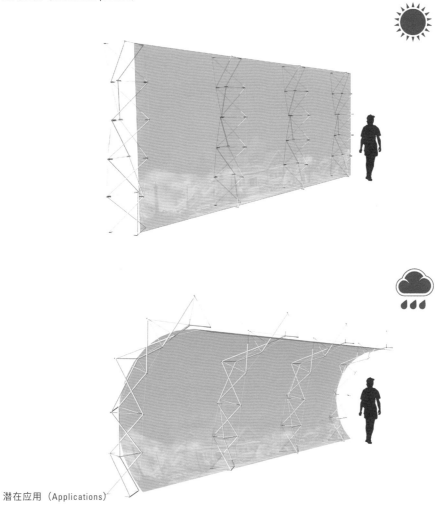

潜在应用（Applications）

04

媒体门 | Media Gate

设计者：2014年互动设计小组：费雨，蔡适然 | 指导教师：虞刚，方立新
Designer: Interactive Design Team 2014: Fei Yu, Cai Shiran | Tutor: Yu Gang, Fang Lixin

"媒体门"试图根据使用者的行为变化，让建筑实体和图像展示都能与使用者产生无缝互动。在"媒体门"中，控制单元主要是识别使用者的位置与动作，同时要让图像展示与门扇开启相互同步，也就是说，当使用者经过时，同时开启相应区域的对应动态图像展示。具体而言，方案设置了一套动态飘雪图像，根据使用者的行为，这套动态图像可以与"媒体门"实体互动部分实现同步运作。"媒体门"在概念设计和原型模型制作过程中，既考虑了机械传动装置的流畅运转，还考虑了传动装置与实体模型以及动态图像之间的整合，最后还考虑了如何创造出迷人的视觉效果。

The Media Gate project attempts to establish seamless interaction between user and media. The controlling unit is able to process the user location and motion while synchronizing the open/close of the door with imagery. In other words, it's projecting the animation on the door whenever a user passes through. In this design, the animation is a snowing vision generated in real time. The snow can also react to user behavior. During the design process, a lot of facts have to be considered, such as the smoothness of the motion, the integration of the real and virtual objects as well as the visual effect of the animation.

设计流程（Design process）

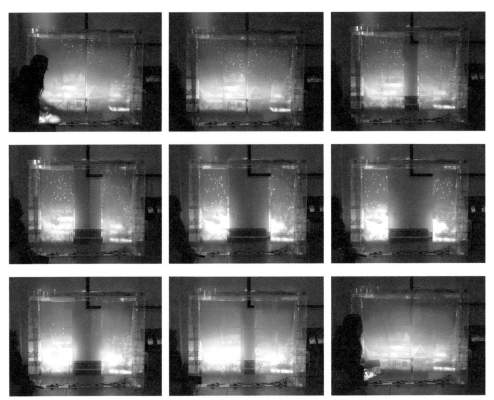

互动过程 (Interaction process)

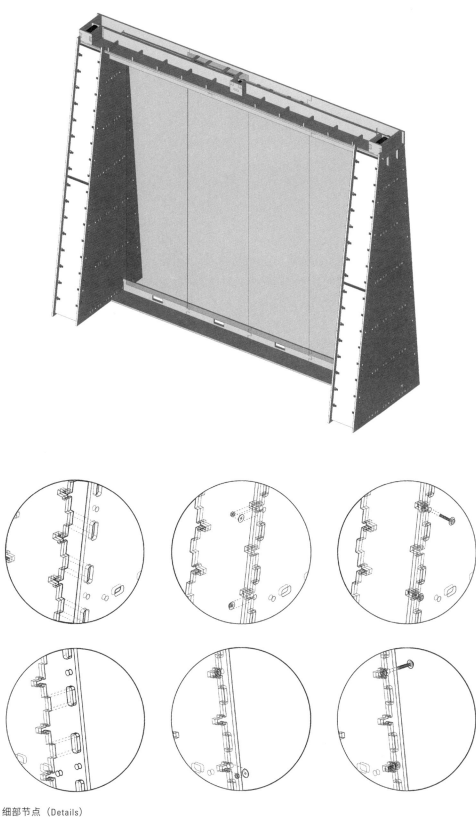

细部节点（Details）

05

动态立面
Dynamic Facade

设计者：郑天宇 | 指导教师：虞刚 | 完成时间：2016 年
Designer: Zheng Tianyu | Tutor: Yu Gang | Complete time: 2016

幕墙系统是目前发展较为成熟的建筑构件系统之一，独立于结构之外，便于和成熟的互动技术产品相结合。而且互动建筑与幕墙的发展轨迹在互动幕墙部分达到契合，互动幕墙也自然成为互动建筑领域的新兴发展方向。"动态立面"试图提出一种双层互动幕墙，前后两层构件可根据光照强度变化而转动不同角度，从而改变建筑遮阳面积以控制进光量，满足不同的采光要求。幕墙构件可根据建筑内部使用者行为而发生改变，当使用者靠近时，其转动速度变慢，且遮挡面积变小以提供更好的景观视野。设计偏重于互动原理的模拟及可能性的探索。

Curtainwall is one of the most developed building component systems, which is free from the building structure and easy to be combined with advanced interactive products. Moreover, the development of interactive architecture and curtainwall has reached synthesis in the topic of interactive curtainwall, which means it will become the new developing direction in the interactive architecture field. The Dynamic Façade attempts to propose a double-skin façade whose components can rotate to different angles according to the intensity of illumination, thus controlling the entering of sunlight. The curtainwall can also interact with the occupants that when they approach the curtainwall, the rotation will slow down and open up for better views. In general, this design focuses on the simulation of interactive principles and the explorations of different possibilities.

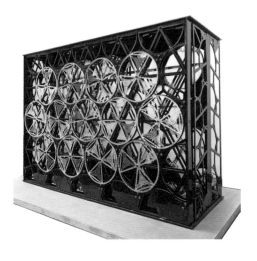

实体模型（Mockup）

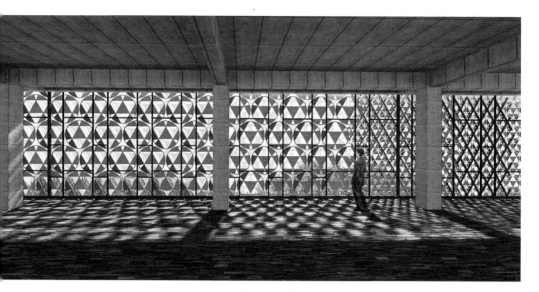

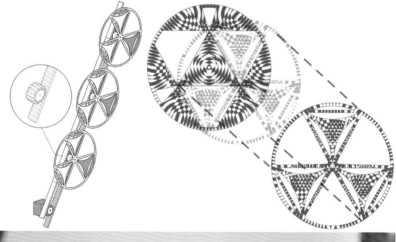

细部节点（Details）

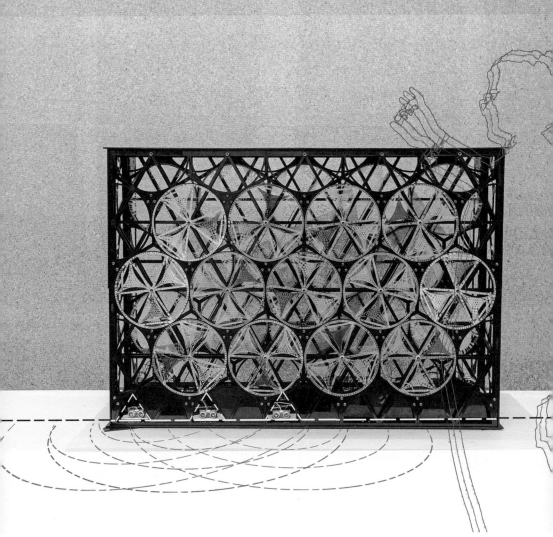

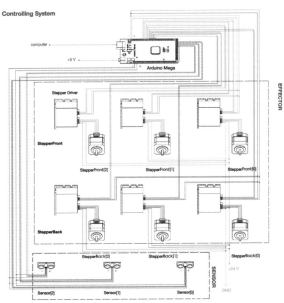

电路连接（Circuit）

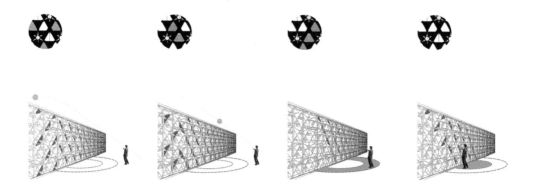

互动过程 (Interaction process)

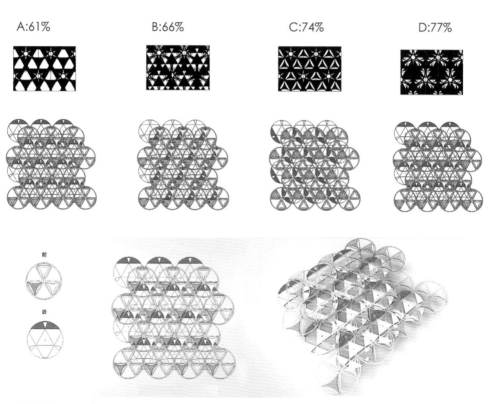

状态组合 (Different states)

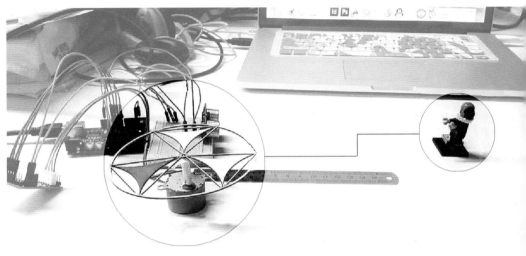

单元构件（Unit）

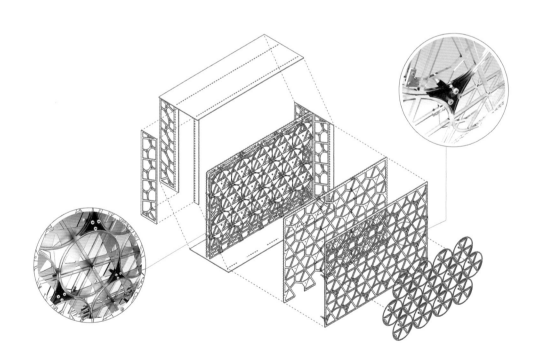

06

弦下
Under the Sine

设计者：2016 年互动设计小组（本科四年级：吕雅蓓 , 王晨 , 徐海闻 , 葛鹏飞）| 指导教师：虞刚 , 李力
完成时间：2016 年

Designer: Interactive Design Team 2016 (Grade4: Lv Yabei, Wang Chen, Xu Haiwen, Ge Pengfei)
Tutor: Yu Gang, Li Li | Complete time: 2016

交互技术在信息时代已经渗透到生活的方方面面，电子元件和机械结构已构成建筑的必要组件，因此，感应使用者行为并做出相应变化的建筑在未来将会成为常态。在此背景下，"弦下"方案希望改变通常的垂直空间限定模式，也就是将常规意义上的"天花板"从静态转换为动态，以满足不同情况中使用者的需求和愿望。在设计和制作过程中，方案先是将"天花板"转换为多个小单元，然后利用感应系统和传动装置将多个小单元的正弦曲线运动模式转化为电机驱动的圆周运动模式，巧妙地实现了多个单元体之间的规则波浪形动态组合。

Interactive technology has been deeply involved into our daily life. Electronic components and mechanical structures are becoming essential parts of building component. Buildings that can adapt to occupants' behaviors are becoming norm in the near future. In such background, the Under the Sine project tries to redefine the vertical space, converting the static ceiling into dynamic according to the demand and anticipation of users. In the designing process, the ceiling is firstly divided into small units. Then a special transmission structure has been introduced to convert the circular motion of step motor into sine wave motion of the unit grid, which greatly reduces the amount of motors and achieves the sine motion combination of multiple units.

设计流程(Design process)

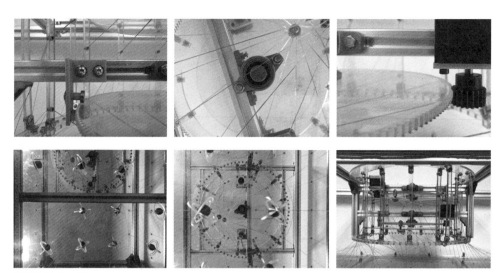

细部节点（Details）

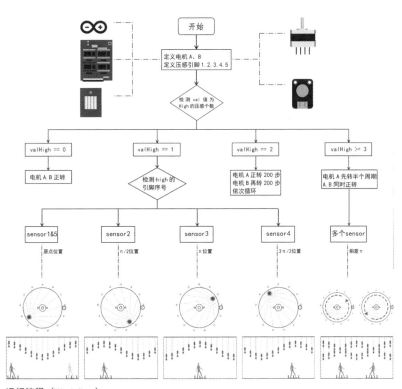

运行流程（Work flow）

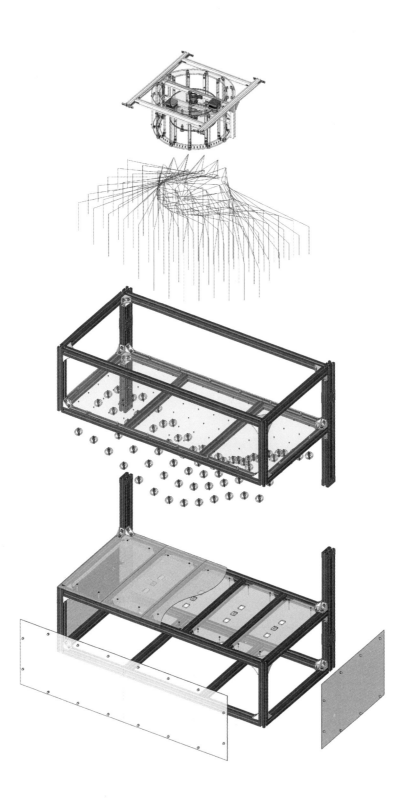

07

叶亭
Leave Pavilion

设计者：2016年互动设计小组（周宇琪，徐思畅，张亚）| 指导教师：虞刚，李力
Designer: Interactive Design Team 2016 (Grade4: Zhou Yuqi, Xu Sichang, Zhang Ya)
Tutor: Yu Gang, Li Li

"叶亭"试图思考建筑物为何互动、为何吸引人和如何吸引人的问题，思考和理解建筑物与群体行为之间的互动关系，以及互动方式对人们日常生活的干预程度。"叶亭"设置了一组可动可展开的独立柱亭，可以根据人们在场所中的集散状况做出相应的变化，一方面为人们提供某种程度的遮风避雨，另一方面也为场所中的人们提供互动的乐趣。根据实际建造方案，这个设计制作了等比例的互动模型，既考虑了电子设备机械装置的整合，又考虑了实际建造的可能性。在实际操作过程中，还对整套传感器和电机进行编码与控制，考虑了零件和材料的组装方式，解决了传动和感应带来的各种摩擦损耗、振动频率和组装问题。

Leave Pavilion is focused on discussing problems such as how can a building interact with people, why and how to attract people's attention and how to be involved in daily life. Leave Pavilion consists of a set of independent foldable umbellar-like structures, which can react according to the distribution of people underneath it. It can provide shelter, as well as entertainment. A mockup, considering the integration of mechanisms and control system, has been built for evaluating the feasibility of real construction. During the building process, some factors, such as the assembly process, friction and vibration are also well addressed.

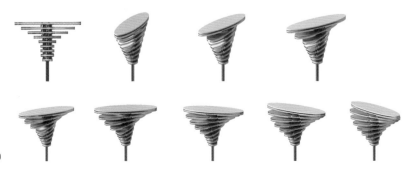

单元构件（Unit）

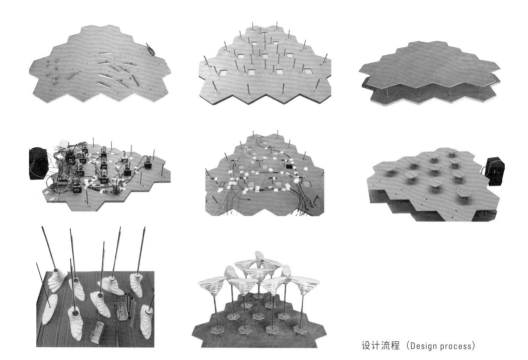

设计流程（Design process）

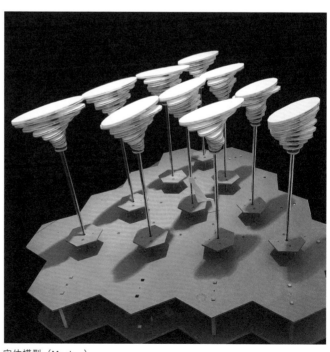

实体模型（Mockup）

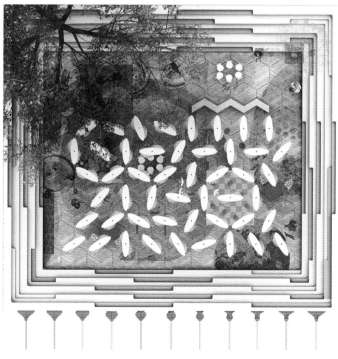

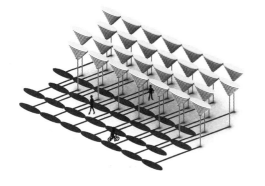

互动过程（Interaction process）

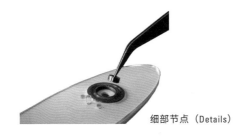

细部节点（Details）

08

动态交织
Kinetic Weaving

设计者：2016年互动设计小组（本科四年级：傅瑞盈，王玥，茆羽）| 指导教师：虞刚，李力
Designer: Interactive Design Team 2016 (Grade4: Fu Ruiying, Wang Yue, Mao Yu)
Tutor: Yu Gang, Li Li

"动态交织"方案希望改变走道或类似空间的封闭属性，试图根据使用者的需要创造动态的空间分割模式，例如廊道模式、半围合模式或斜线穿越模式，以便改变通常的水平空间划分模式。此设计搭建了等比例模型，以便模拟几种空间模式的开启和闭合。这个方案的主体空间由12个板块单元围合而成，每个模块由一个步进电机在水平方向驱动，以形成不同的围合和空间限定方式。与人互动的传感器分别安装在地板和侧面，12个板块单元会根据感应装置对使用者行为判断，同时做出相应的空间划分组合和变化。

The Kinetic Weaving project tries to open up the corridor space by giving the wall a mobile property. Space can be divided into different forms such as full enclosure and half enclosure according to different demands. A mockup has been built to simulate different states of openness. It consists of 12 wall units, and each of them is controlled by a step motor. Sensors for occupancy detection are installed underneath the floor tiles. Sensor will be triggered when people step onto it and 12 units will work collaboratively to divide the space.

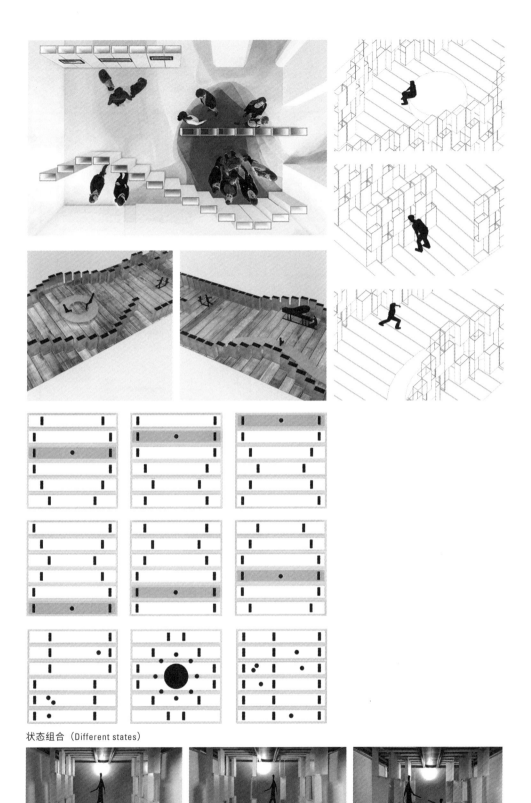

状态组合 (Different states)

互动过程 (Interaction process)

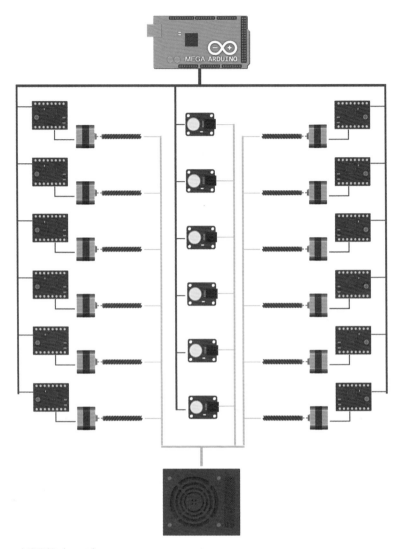

电路连接（Circuit）

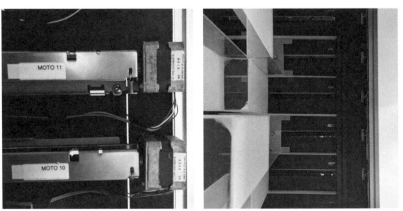

细部节点（Details）

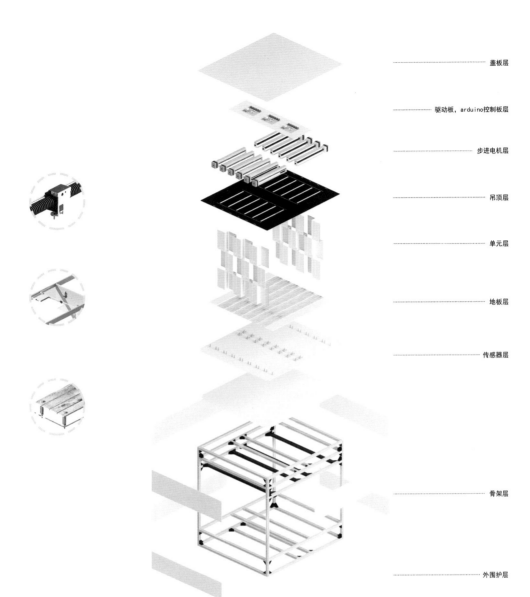

09

信息墙
Hexagon Info-Wall

设计者：2016年互动设计小组（本科四年级：邹传正，蒋钰人，方志华）| 指导教师：虞刚，李力
Designer: Interactive Design Team 2016 (Grade4: Zou Chuanzheng, Jiang Yuren, Fang Zhihua)
Tutor: Yu Gang, Li Li

针对相对单一或枯燥的建筑室外空间，以及现有城市公共空间中信息展示的无序和混乱状态，"信息墙"试图设计和搭建一个包含六边形"像素"的信息展示墙，为公众提供具有参与性和互动性的公共场所，同时激活建筑、场地和使用者，并让三者之间产生互动。"信息墙"设计不仅展开了一系列可行的试验性设想，还搭建了一个等比例的局部单元模型，既考虑了"墙面"的视觉审美效果，又考虑了整合电子机械设备的各种可能性，以便模拟真实情况下使用者与"信息墙"之间的互动，为真实建造提供可行和合理的参考。

The Hexagon info-wall is an interactive media wall consists of several hexagon units. It aims to create interactive public space that can replace the indifferent outdoor space and deal with the rambling information display condition. It is also supposed to activate the interactions among building, field and people. Not only a series of experimental ideas are brought up, a full size unit mockup is also built for evaluating possible combination and aesthetic effect to simulate the interaction between user and info-wall as well as providing feasible and reasonable reference for construction.

设计流程（Design process）

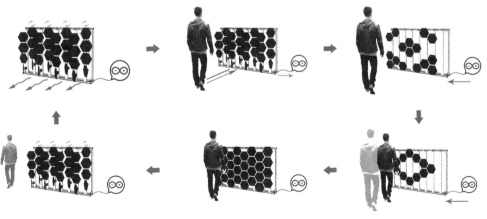

互动过程（Interaction process）

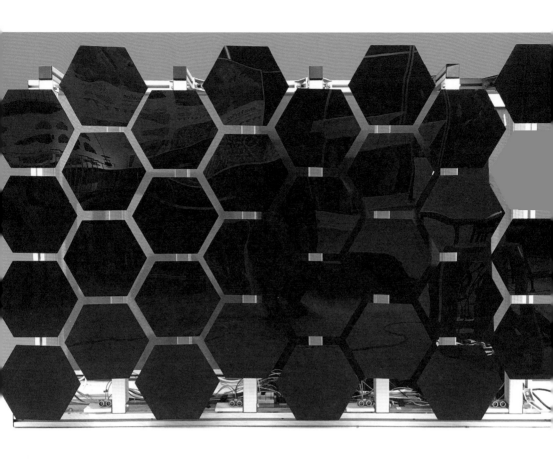

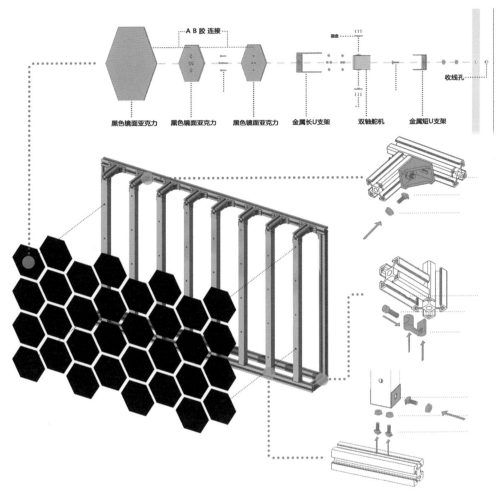

细部节点（Details）

A Self-Organizing Wireless Sensor Network for Indoor Environment Surveillance

李力

原文刊载于 DADA2012 国际会议论文集。

Research Background

With the development of the computer, internet, and ubiquitous computing technology(Krumm 2009), the technical difficulty and cost of data acquisition, storage, and calculation have been increasingly reducing. Through the employment of these technologies, researchers can acquire a large number of data available for analysis from the natural environment, urban environment, internet, and financial market. By means of data mining and construction of the correlation among the data, researchers can carry out corresponding analysis and predication of the study objects. By replacing the original model of sampling investigation with the Big Data and replacing the search for the causality among events with the discovery of the correlation between data(Mayer-Schonberger and Cukier 2013), researchers are able to make easier predications and find out the potential relevance among events.

Due to the increase of data demand, wireless sensor network, as a tool of data acquisition, has been applied more and more widely. Wireless sensor network refers to distributed sensors linked by wireless media. Because the transmission distance of a single wireless transceiver is limited, the information transmission is usually realized by the method of multi hop(Akyildiz, Wang et al. 2005). Compared to the traditional wired network, wireless sensor network possesses better flexibility and scalability. At first, the network was used in the military field, particularly battlefield monitoring, and now it has been gradually applied to forest, wild animal, factory area, and indoor environment monitoring(Chee-Yee and Kumar 2003).

As for the monitoring of building indoor environment, a relatively mature building automation system has been formed in the public building field while the monitoring of residential building indoor environment still stays on the experimental stage. Besides privacy reasons, several factors are involved in the technological aspect(Edwards and Grinter 2001, Eckl and MacWilliams 2009):

- First, the operation of the control system is too complicated to make the system settings as intuitive as the light switch, but at an average home, it is not likely to have a particular person responsible for the automation system like that in the public building field. In fact, a user needs to undergo a certain learning curve in order to learn and adapt to the system.

- Second, due to different durability of various buildings, the installing and updating of control systems, especially wired control systems, will inevitably cause certain damages to the original house structure, decoration, and devices, which will also need professional operation.
- Last but not least is the price problem which includes not only the price of monitoring equipment, but also installation and maintenance expenses, and the energy consumption of the system itself.

At present, different research institutes approach from different aspects to make optimizations of the wireless sensor network. For example the research from (Surie, Laguionie et al. 2008) focused on the signal strength in real house environment. The AlarmNet (Wood, Stankovic et al. 2008) discussed about the context-awareness ability. Berkeley Motes (Hill and Culler 2002) and Telos (Polastre, Szewczyk et al. 2005) are designed for low-cost usage. In contrast, this study focused on the research of the practicality and stability.

Research Purposes

By integrating the wireless sensor network, ubiquitous computing, and database technology, this study intended to build a control system designed for the residential environment. At the same time, the system will possess relatively high adaptability, ease of use, openness, and stability, and it should be suitable for amateurs without much electronics and programming expertise and architecture major researchers. The system can be integrated in the Smart Home control system, with the following features.

- Self-organization: In the network aspect, the wireless mesh network will be formed automatically after the system is powered. The network topological structure can update automatically with the position changes and increase and decrease in quantity of the sensor nodes. In the database aspect, according to the composition of sensors in the network, the database will automatically generate corresponding database structures.
- Easy deployment: By means of the wireless network, it can avoid the wiring difficulties of wired network. Wireless mesh network demands few restrictions of sensor positions, thus greatly increasing the freedom of sensor deployment.
- Compatibility: Through the gateway node, it can communicate with different network protocols like Wi-Fi and Bluetooth devices. By storing data into the universal database, the system makes it convenient to exchange data between different programs and conduct secondary development.
- Low energy consumption: The adoption of low energy-consuming wireless transceiver and low-rate micro controller will reduce the energy consumption of system operation.
- Independence: The system is capable of data acquisition and logging without relying on other computer and internet connections.
- Self-healing: If a certain sensor node in the system operates abnormally, the node will automatically detach itself from the network and the network will update the topological structure to ensure normal data transmission. The detached node will automatically join the network after returning to normal.

System Introduction

The whole system is composed of three parts: hardware, software, and communicating protocol. In the hardware, every sensor node in the system will contain a micro controller and wireless transceiver module. The micro controller will receive and process the information acquired by all the sensors and the packaging and analysis of wireless data. The system adopts ATmega328P micro controller on Arduino Pro Mini development board and the wireless transceiver module is used to form the wireless mesh network and send and receive wireless data and instructions, adopting Xbee DigiMesh 2.4 Wireless RF Module.

The ATmega328P micro controller is an 8-bit AVR micro controller with the 32kb readable and writable flash rom, plus 1kb eeprom and 2k SRAM(Atmel 2013). On the Aduino Pro Mini development board there are 13 digital reading and writing ports, six 10-bit A/D converters input ports, and a 16 Mhz crystal oscillator which is set up for the micro controller. Aduino Pro Mini also supports SPI, I2C, and serial communication at the same time.

Xbee DigiMesh 2.4 Wireless RF Module works in the 2.4G band. It adopts the DigiMesh communicating protocol, and it can form mesh networks automatically and realize dynamic peer-to-peer communications. Besides, it also supports the broadcast mode and multi-cast mode, with a transmission power of only 1 mw, a transmission distance about 90m without obstacles, and a highest transmission rate of 250kb/s. The module exchanges data with the micro controller through serial ports and the data exchange mode includes the transparent mode and API mode(Digi 2013). The system adopts the API mode with every wireless module possessing a unique 64-bit communication address. Another Xbee ZB Wireless RF Module can also be applied to this system. It adopts the Zigbee transport protocol and has better compatibility than Digimesh but a more complicated network structure(Digi 2013).

According to different wireless node functions, the nodes can be classified into data logger nodes [Figure 1] and sensor nodes [Figure 2] with the operating voltage of 5v. A network system incorporates one data logger node and several sensor nodes. The data logger node is also called gate way or sink node(Kazem Sohraby, Daniel Minoli et al. 2007). It is mainly used to record submitted data by sensor nodes. Moreover, it is also responsible for the management of other nodes in the network, including node discovery, adding, and deletion. When an abnormal event occurs

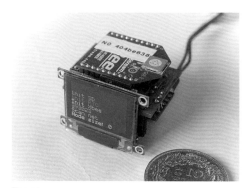

Figure 1
Data logger node.

Figure 2
Sensor node.

at the data exchange model point or during data transmission, the node has to record the abnormal event into the log file. When necessary, it also needs to submit the data to the data manager software. Apart from the micro controller and wireless transceiver, the data logger node is also furnished with the real-time clock module which provides the submission time of every sensor data, the SD memory card which records sensor data and log files, and the LED display which shows the operating state of the network. The real-time clock module adopts the DS1307 chip and provides year, month, day, hour, minute, and second data. The module is connected to the micro controller with the I2C protocol. The SD card adopts the 512M mini SD card and connected to the micro controller with the SPI protocol. The LED display adopts the 96*64 OLED and connected to the micro controller with the serial protocol.

The sensor node is equipped with sensors of many kinds and used to collect sensor data and send them to the data logger node. There are also two kinds of sensor nodes: the passive node and active node. The passive node will send data after it receives upload instructions. It is used to monitor continuously changing information like temperature and humidity. The active node will send data when specific events occur and it is used to monitor jump changing events like opening and closing of doors and windows. This design can, on one hand, acquire data with slow continuous change and reduce data transmission and its energy consumption by certain sampling frequency, and on the other hand, capture instantaneous events occurring at sampling intervals. Take this experiment for example; the passive node is furnished with temperature and humidity, luminous intensity, sound, and human infrared sensors. Among them, the temperature and humidity sensor adopts the MTH02A single-interface digital chip and luminous intensity sensor adopts the BH-1750FV1 digital sensor and is connected by I2C. The PIR sensor adopts the RE200B passive infrared sensor, and the sound sensor adopts the BCM-9765P electronic microphone. By contrast, the active node is only furnished with a reed switch sensor to control the opening and closing of doors and windows.

The software can be divided into three parts: the in-chip control program within the micro controller, remote programmer software [Figure 3] in the desktop system, and data management software [Figure 4]. The in-chip control program drives the micro controller and the data exchange with connected sensors, and

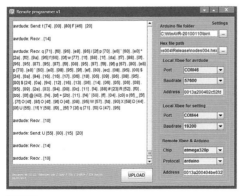

Figure 3
The interface of remote programmer software.

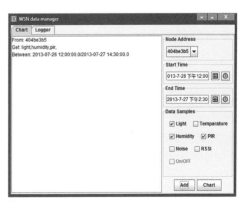

Figure 4
The interface of data management software.

controls the working flow of nodes. It is developed by the c++ language and downloaded into the flash rom of the micro controller after being compiled. The remote programmer software downloads compiled control programs into the micro controller through wireless ways. This software is written in Java and operated in Windows desktop systems. Through the Xbee wireless transceiver which is connected to computer serial ports, the software wirelessly transmits programs to the Xbee wireless transceiver of the sensor nodes, and then through Xbee serial communication writes them into the flash rom of the micro controller. Compared to the traditional method of wired download, it avoids the trouble of wiring and provides great convenience for updating installed wireless node in-chip programs. The data management software reads the raw data stored in the data logger node SD card, transforms the data in certain methods, and saves them into the database with database management software. Meanwhile, it can also read the historical data from the database as it is demanded and present them in graphical forms. The software is written in Java and operated in Windows desktop systems. The database management software adopts MySQL which provides programming interfaces in many languages and suits secondary development of researchers. The database management software can be installed in personal computers and the network server as well, thus enabling researchers to share and manage data conveniently.

Communicating protocol is used to control the instruction sending and data exchange between the nodes. The protocol is nested in the application layer protocol provided by Xbee and composed of instruction names and parameters. The main instructions include adding nodes, deleting nodes, collecting data, inserting data, joining network, and exiting network, etc, and there are some special instructions including LED switch instructions, date change instructions, and pause instructions.

Work flow

Based on the thought of non configuration, the system operation is automatically completed by the in-chip program of every sensor node, without relying on external operations [Figure 5]. Nodes with different functions own two states which are respectively the initialization and loop state. The initialization process is used to initialize the micro controller and connected sensors and equipment, discover networks, and record the 64-bit addresses of the nodes. When the initialization ends, the nodes will enter the loop state in which they constantly update the sensor data, monitor wireless network data, and execute wireless instructions.

Work flows of different nodes vary and that of the data logging node is as follows:
- Connect to the power, initialize all devices, and acquire current date and time.
- The Xbee wireless transceiver automatically forms the network, and joins the wireless mesh network.
- Scan the network, record the number of found sensor nodes, and acquire the 64-bit addresses of all nodes. Write them into eeprom if their addresses are not in it.
- Start the timer, and enter the loop state after the initialization ends.
- Monitor Xbee, and analyze the 64-bit addresses and instruction contents of the source nodes once receiving instructions. If it is the instruction of joining nodes, reply the acknowledge instruction and add the

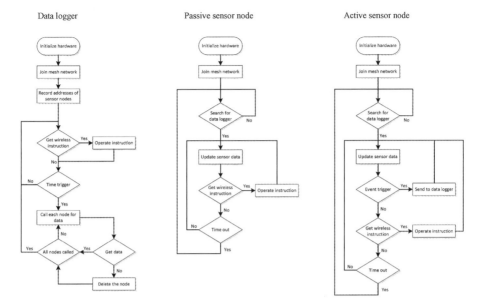

Figure 5
Diagrams of the work flows of different nodes.

64-bit addresses into eeprom. If it is the instruction of inserting data, record the data into the SD card.
- Check the timer. If it arrives at the specified time, start to collect data from the nodes, extract the 64-bit addresses one by one from eeprom, and send the upload data instruction to this node. After the target node returns data, record them into the SD card if receives the data. If the target node fails to return data for three consecutive times, then delete its node address from eeprom.
- Reset the timer and prepare for the re-loop.

The work flow of the passive sensor node is:
- Connect to the power and initialize all devices.
- The Xbee wireless transceiver automatically forms the network, and joins the wireless mesh network.
- Send the instruction of joining nodes in the broadcast mode and wait for the reply from the data logging node. If it receives the reply, then record its 64-bit address.
- nitialize the timer and enter the loop state.
- If the timer time-out occurs, restart the node.
- Monitor the wireless data received by Xbee. If it receives the upload data instruction, read the data of every sensor and send them to the data logging node. Reset the timer.

The work flow of the active sensor node is:
- Connect to the power and initialize all devices.
- The Xbee wireless transceiver automatically forms the network, and joins the wireless mesh network.
- Send the instruction of joining nodes in the broadcast mode and wait for the reply from the data logging node. If it receives the reply, then record its 64-bit address.

- Initialize the timer and enter the loop state.
- If the timer time-out occurs, restart the node.
- Monitor the sensor. If jump change occurs, send the instruction of inserting data to the data logging node and send the data as well.
- Monitor the wireless data received by Xbee. If it receives the upload data instruction, read the data of every sensor and send them to the data logging node. Reset the timer.

We can find that the operation of every node is relatively independent. The sensor node can either be passively discovered by the data logging node or join the system actively; therefore, the starting process of the whole system is not in a specified order. The 64-bit addresses of the sensors do not have to be preset; they can be discovered and saved during operation. The nodes can be added and deleted at any time, which possess high flexibility. The data logging node can automatically remove the sensor nodes with abnormal events based on the success rate of the data transmission, thus preventing ineffective data transmissions. Also the sensor nodes will automatically restart themselves after noticing abnormal events of their own and try to rejoin the network, which ensures that the network possesses certain self-healing abilities.

System Evaluation

To detect the actual operational effect, stability, and energy consumption of the system, the system was installed in real residential environment which was a student apartment about 12m^2 [Figure 6] . Altogether eight nodes were installed in the room, including one data logging node, five passive sensor nodes, and two active sensor nodes. Two of the passive sensor nodes were placed outdoors to monitor the outdoor and corridor environment respectively. The remaining three were distributed uniformly in the room [Figure 7] . The active sensor nodes were respectively installed on the window frame and door frame to monitor the opening and closing of the windows and doors.

The data logging node collected data from each sensor node every ten seconds and the data volume which the data logging node recorded every day amounted to 1M bytes. Every passive node produced about 170K bytes and every active node produced about 80k bytes. Plus the space occupied by the log file, a 512M SD card was enough to store a year's data volume.

Different kinds of graphical time series chart, which can give the user more intuitive feelings of the data, can be created easily by the data management software. User can combine different sensor data or sensor data from different nodes into one multiple axes time series chat to find out the correlation among the data [Figure 8] . User can also create the chart in different time scales to discover the daily or seasonal change [Figure 9] .

The collected data were saved into the database through the data management software. Data of different nodes were separately saved in different tables which were named by the lower 32 -bit address of the node. Every row in the table saved a piece of information sent by the node and the first column saved the time stamp, recording the sending time of the information. Other columns respectively saved the numerical values returned by every sensor [Figure 10] .

As for the stability of wireless data transmission, nodes with best signals like the indoor

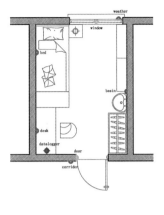

Figure 6
The plan of the room. The blue, red and green points represent data logger, passive node and active node respectively.

Figure 7
The sensor node which is placed close to the bed.

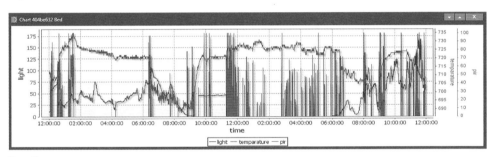

Figure 8
The chart created by the data management software: one day's sensor data collected by one sensor node, includes light intensity, temperature and movement detection data.

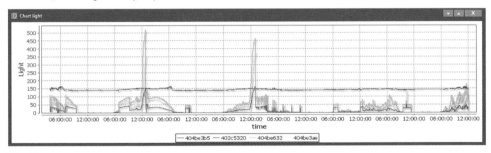

Figure 9
The chart created by the data management software: four days' light intensity data from four different sensor node

nodes could keep the range within -60dbm~-65dbm. Even if the nodes receive disturbance of other electromagnetic signals, they could still ensure the received signal intensity more than the lowest receive power of Xbee, namely -95dbm, without losing data packages. The nodes in the corridor, because of the wall blocking it from the data logging node, kept the signal intensity between -76dbm~-83dbm under normal circumstances [Figure 11]. When being interfered, they lost 2% package of the total number, which did not create great influence on the data integrity.

In the aspect of energy consumption, be-

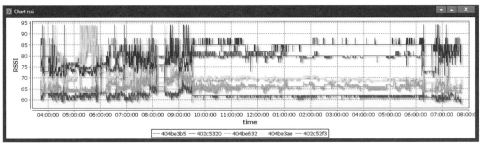

Figure 10
The data structure inside the database

Figure 11
The chart created by the data management software: RSSI of different sensor nodes

cause the data logging node shouldered the most work intensity, its peak current magnitude amounted to 500mA under the 5v voltage. There were more sensors on the passive nodes, so their peak current magnitude reached 180mA, and that of the active nodes was 60mA. If we use the common charging battery pack of 2400mAh to power them, they can respectively sustain about 5, 13, and 40 hours.

Existing problems and follow-up studies

From the test results we can see that the whole control system basically reached the expected effects. But in the aspect of energy consumption, although we made optimizations through the methods of less data transmission and adopting low-rate micro controller, etc, the present battery technology was not adequate to support long operation of the system and each node still had to rely on the wired power supply mode of the power adapter. This also caused some limitations of the node deployment. The solution of this problem can be realized by more use of the sleep-waked mode through selecting micro controllers and sensors with lower energy consumption and optimizing the work flows of the nodes. Moreover, since we still stayed in the prototype design stage, we have not made deeper research and design of the size, appearance, and package of the nodes. In the follow-up design we will take the appearance influence on installment and indoor environment into consideration.

Through the system, researcher can conveniently acquire continuous and high density indoor environment data. Based on this, we can study the relevance between data and resident behavior patterns, between data and data, and provide necessary data support to improve the building performance and the residence intelligence degree.

REFERENCES

Akyildiz, I. F., X. Wang and W. Wang (2005). "Wireless mesh networks: a survey." Computer Networks 47(4): 445-487.

Atmel (2013). ATmega328P data sheet.

Chee-Yee, C. and S. P. Kumar (2003). "Sensor networks: evolution, opportunities, and challenges." Proceedings of the IEEE 91(8): 1247-1256.

Digi (2013). Wireless Mesh Networking ZigBee® vs. DigiMesh™.

Digi. (2013). "XBee® DigiMesh® 2.4 ", from http://www.digi.com/products/wireless-wired-embedded-solutions/zigbee-rf-modules/zigbee-mesh-module/

xbee-digimesh-2-4.

Eckl, R. and A. MacWilliams (2009). "Smart Home Challenges and Approaches to Solve Them: A Practical Industrial Perspective." Intelligent Interactive Assistance and Mobile Multimedia Computing 53: 119-130.

Edwards, W. K. and R. E. Grinter (2001). At home with ubiquitous computing: seven challenges. Ubicomp 2001: Ubiquitous Computing, Springer.

Hill, J. L. and D. E. Culler (2002). "Mica: A wireless platform for deeply embedded networks." Micro, IEEE 22(6): 12-24.

Kazem Sohraby, Daniel Minoli and T. Znati (2007). Wireless Sensor Networks: Technology, Protocols, and Applications Wiley-Interscience.

Krumm, J. (2009). Ubiquitous Computing, Taylor & Francis.

Mayer-Schonberger, V. and K. N. Cukier (2013). Big Data: A Revolution That Will Transform How We Live, Work, and Think, Eamon Dolan/Houghton Mifflin Har-

court.

Polastre, J., R. Szewczyk and D. Culler (2005). Telos: enabling ultra-low power wireless research. Information Processing in Sensor Networks, 2005. IPSN 2005. Fourth International Symposium on, IEEE.

Surie, D., O. Laguionie and T. Pederson (2008). Wireless sensor networking of everyday objects in a smart home environment. Intelligent Sensors, Sensor Networks and Information Processing, 2008. ISSNIP 2008. International Confer-

ence on, IEEE.

Wood, A., J. Stankovic, G. Virone, L. Selavo, Z. He, Q. Cao, T. Doan, Y. Wu, L. Fang and R. Stoleru (2008). "Context-aware wireless sensor networks for assisted living and residential monitoring." Network, IEEE 22(4): 26-33.

Sequential Behavior Pattern Discovery with Frequent Episode Mining and Wireless Sensor Network

李力

Abstract—By recognizing patterns in occupants' daily activities, building systems are able to optimize and personalize their services. Established technologies are available for data collection and pattern mining, but they all share the drawback that the methodology used for data collection tends to be ill suited for the purposes of pattern recognition. For this research, we have developed a bespoke Wireless Sensor Network (WSN) and combined it with a compact data format for Frequent Episode Mining (FEM) to overcome this obstacle. This proposed framework has been evaluated with both synthetic data from a smart home simulator and with real data from a self-organizing WSN in a student home. We have been able to demonstrate that the framework is capable of discovering sequential patterns in heterogeneous sensor data. With corresponding scenario, patterns in daily activities can be deduced. The framework is self-contained, scaleable and energy efficient, and is thus applicable in different building system settings.

Index Terms—Smart City, Smart Building, Frequent Episode Mining, Wireless Sensor Network

原文刊载于 IEEE Communications Magazine 2017 年 7 月刊。

I. Introduction

From a technical point of view, smart city tries to improve the life quality of its citizen in terms of urban services by utilizing Information and Communication Technology (ICT), Data Mining (DM), and other new technologies. People's daily behavior is a decisive factor of a lot of aspects of urban environment, such as traffic, air quality, energy cost, etc. Considering people spend about 90 percent of their lives indoor, it's no surprise that buildings are responsible for about two-thirds of all electrical energy consumption. So, making cities smarter often begins indoors. By enabling data-driven decisions for smarter buildings, utilizing data collection to reveal occupants' behavior patterns, it is able to anticipate the usage of the buildings, thus reduce the consumption while improve the experience[1].

The data-driven decision for buildings is largely based on data acquisition and data mining technologies. Wireless Sensor Networks are widely used for data acquisition nowadays, since it's low cost, and more flexible than wired solutions. Plenty of researches have been done recently in efficiency [2][3], and mobility [4-6]. Since it is impossible for the system designer to envision all possible contexts forehead, decision making in control system largely relies on data mining and machine learning techniques, such as Artificial Neural Network (ANN), Support Vector Machine (SVM), Self-organizing Map (SOM), Hidden Markov Model (HMM), and Frequent Pattern Mining (FPM)[7].

A. Existing problem

A lot of studies have been conducted on data acquisition and data mining for the building system, however, no practical solution has been provided yet. There are several reasons:

- Too complex. Most of the researches were The conducted in experimental environment. The installation, maintenance, and upgrade of the systems need expertise. Some mining algorithms require a lot of parameters settings workings. Those parameters are not very intuitive and need professional and prior knowledge of the environment. The supervise algorithms need additional training data that are hard to get in real world application.
- Too simply. From building system, there are mainly two kinds of data types can be recorded: numerical (discrete and continuous sensor values), categorical (weather condition: windy, snowy, sunny, etc.). Unfortunately, most of the algorithms only can deal with only one type of them. Although the datatype are interchangeable, but additional parameters and prior knowledge of the dataset are required.
- Gap in research fields. Established technologies are available for data collection and data mining, but they all share the drawback that the methodology used for data collection tends to be ill suited for the purposes of data mining. The WSN developers keep improving the efficiency, regardless what kind of data and measuring frequency are actually needed. The DM researchers focus on the accuracy of the algorithm, regardless if the data source is available or can be collected efficiently.

B. Solution

The research attempts to integrate data acquisition and data mining techniques more efficiently and practically for discovering behavior patterns. To achieve this, we first proposed a compact data format encompasses both sensor data about spontaneous events and periodic environmental readings. With this format, less data set and transmission are required from WSN, while the environmental information can be used to reduce the redundancy and deduce behavior patterns. For data acquisition, a mesh WSN called Self-organizing WSN is designed, which requires neither planning nor configuration. Frequent Episode Mining (FEM) algorithm is selected for mining the sequential patterns. It is adapted to mine on both categorical and numerical data in the dataset by introducing the DBSCAN clustering algorithm. Finally, a framework including is proposed to combine all these technologies seamless. It is evaluated with both synthetic data from a smart home simulator and with real data from a self-organizing WSN in a student home. We have been able to demonstrate that the framework is capable of discovering sequential patterns in heterogeneous sensor data. With corresponding scenario, patterns in daily activities can be deduced. The framework is self-contained, scaleable and energy efficient, and is thus applicable in different building system settings.

II. The compact data format

The standard building sensor data format hasn't established yet. There are mainly two ways for data recording: event based and interval based. The event bases approach only records when the sensor is trigged. The collected data size is small, but some slowly changing environment parameter will be missed. The interval based approach do the sampling under a fixed rate. It has to increase the frequency to avoid missing short event, but the data size will grow correspondingly. In our approach, we mix these two approaches. The building sensor data is divided into two categories event data and ambient data. The event data means the sensor records that triggered by the occupant, such as open/close the door, turn on/off the light. The ambient data refer to the environment parameters, such as temperature, light intensity. It provides the context and the scenario for the events. Here, time is also regarded as ambient data. With this approach, we keep the sampling rate as low as possible without missing any event.

III. The data acquisition

The Self-organizing WSN is a mesh network designed to minimize the settings and configurations and dependence on infrastructure during the installation, so it can be easily deployed or removed. With the self-organizing ability, the network can form the mesh network automatically when network is installed or new node be added.

A. Sensor Nodes

The network consists of three types of node: gateway node, active sensor node, and passive sensor node. Each sensor nodes is equipped with an ATmega328P micro controller for processing and XBee DigiMesh 2.4 Wireless RF Module for communication. Besides that, additional components are added according to the node's task.

The gateway node is mainly used to record data collected from the network. Moreover, it is also responsible for the management of network operation, including node discovery, adding, deleting, and error control. It is equipped with a DS1307 real-time clock module which provides the time stamp for each sample, It use a SD card for data and log file storage, and a LCD for displaying real time information (Figure 1, a).

The passive node is used to collect ambient data and will send data after receiving an upload request from the gateway node. The active node is designed for detect event. It will send data whenever specific events occur. This design can, on one hand, reduce data transmission and energy consumption by running on a relative low sampling rate, and on the other hand, capture instantaneous events that occur in sampling intervals. In this experiment, the passive node is equipped with temperature, humidity, luminous intensity, and passive infrared (PIR) sensors (Figure 1, b). By contrast, the

Figure 1. Sensor nodes: a, gateway node, b, passive sensor node

active node is only equipped with a reed switch to detect the opening and closing of doors and windows.

B. Database

A data managing program is provided, which is a JAVA program communicates with MySQL database. It parses the raw data stored on gateway node, and separates them into according table of each sensor node that are named by the low 32-bit address of the Xbee address. In the table, the first column is the timestamp of the sample, and the rest columns respectively save sensor values. Meanwhile, the database managing program can also read the historical data from the database on demand and visualize the data in different time series chart. This kind of chart can help to provide an intuitive understanding of the correlations between different sensor data.

IV. Pattern mining

The proposed algorithm is trying to provide a data mining technique, which requires less parameter, non priori knowledge, and provide more meaningful sequential pattern. FEM was selected as core algorithm. The advantages of using FEM algorithm are: (1) It needs only one simple parameter, i.e. minimum support. (2) It allows gaps in the pattern, which gives it certain tolerance to the randomness and noise in the real life data. (3) The mining process is unsupervised, no training data is needed, and no pre segmentation is needed. (4) There is no randomness in output patterns.

However, it also has some disadvantages in mining building sensor data: (1) It only works with categorical data. (2) There could be a lot of redundancy and meaningless patterns in the mining result. In this research, a preprocessing module is introduced to convert numerical ambient sensor data into categorical data without priori knowledges. By introducing the associated ambient sensor data and complex temporal database, the redundancy can be significantly reduced. With environment information in the mining result, patterns in daily activities can be deduced.

Meanwhile, by introducing more complex data structure and visualization module, it is also possible to display occurrences of the patterns and mining process.

A. Preliminary

In 1993, Agrawal invented the Apriori algorithm to find all co-occurrence relationships, called associations, among data items [8]. He first applied downward-closure property in finding frequent itemsets in the transaction database. This property narrows down the search space drastically and enables the mining on large scale database.

FEM is one of Apriori based algorithm mining with temporal database. For example: 〈(2 4)(1 3 4)(6)(5)(3 7)(6)(2 3)(1 3 6)(5 8)(7)(6)(2)(1 2 3)(4 5)〉 is a temporal database. Each number inside is called an item. Numbers inside in one pair of brackets form an itemset. The position of itemset in the database indicates the sequence of their appearances. Item in the same itemset appeared at the same time. Given a minimum support 3, <(2)(1 3)(5)> is one of the frequent episodes, because it appears ≥ 3 time in the database. Similarly, <(6)>:4 and <(2)(1 3)>:3 are also frequent episodes (the value after colon represents its amount of appearance in the database). <(2)(1 3)(5)> and <(6)> are called closed frequent episode, because they has no super sequence has same frequency with them. <(2)(1 3)> is not closed, because <(2)(1 3)(5)> have the same frequency. More formal definition can be found in [9]. In short, our algorithm is able to extract closed frequent episodes on the complex temporal database consists of heterogeneous sensor data types.

B. Preprocessing module

Both numerical and categorical sensor data have to be converted into categorical values (i.e. item) for the temporal database. In this case, natural numbers are used to represent the item ID. It is not difficult to assign categorical values with item ID. For example the on/off of one light can be assigned with "1" and "2" respectively. However, for the numerical value, such as the room temperature, ranging from 5 to 30 degrees, it is impossible to assign a symbol for each value recoded. Moreover, only the ambient values at the time point when some event happens are interested. Such ambient data called associated ambient sensor data, for example, the temperature when a certain window is opened.

People usually perform their daily activities at similar times and in similar circumstances.

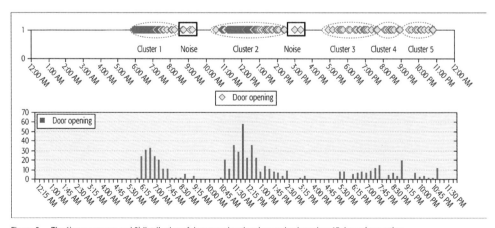

Figure 2. The 1)occurrences and 2)distribution of door opening time in one day based on 15 days observation

This assumption can be demonstrated by the data collected for a real life experiment In Figure 2, there is an overlap of 15 days door opening event into one day. It is clear that there are several high density time regions.

DBSCAN [10] is chosen for clustering the associated ambient sensor data. The reason is: (1) It is a density-based clustering algorithm which fits our need. (2) it needs one one parameter, max distance. And this value can be calculated based on the historical data. (3) It recognizes isolated data point as noise rather than tries to cluster them.

The creation of the temporal database consists of 3 main steps, Figure 3 shows the whole process:

First, each event sensor data is assigned with an item ID. For example, window open

a, sensor data tables

1 Window sensor		3 temperature sensor data	
Time stamp	state	Time stamp	value
08/18 6:39 a.m.	open	08/18 6:39 a.m.	27.5
08/18 5:21 p.m.	close	08/18 7:30 a.m.	28.8
08/19 7:24 a.m.	open	08/18 4:22 p.m.	19.9
08/19 8:14 p.m.	close	08/18 5:21 p.m.	18.3
08/20 8:05 a.m.	open	08/19 7:24 a.m.	26.8
08/20 10:09 p.m.	close	08/19 8:15 a.m.	23.7
08/21 6:33 a.m.	open	08/19 4:16 p.m.	21.9
08/21 9:36 p.m.	close	08/19 8:14 p.m.	17.5
2 Door sensor		08/20 8:05 a.m.	28.4
Time stamp	state	08/20 8:45 a.m.	25.5
08/18 7:30 a.m.	open	08/20 4:19 p.m.	12.4
08/18 4:22 p.m.	close	08/20 10:09 p.m.	22.9
08/19 8:15 a.m.	open	08/21 6:33 a.m.	22.5
08/19 4:16 p.m.	close	08/21 7:50 a.m.	20.4
08/20 8:45 a.m.	open	08/21 4:26 p.m.	15.7
08/20 4:19 p.m.	close	08/21 9:36 p.m.	22.2
08/21 7:50 a.m.	open		
08/21 4:26 p.m.	close		

b, item ID mapping table

Events and clusters	Item ID
Window open	1
Window close	2
Door open	3
Door close	4
Cluster (27.5°C, 26.8°C, 28.4°C)	5
Cluster (18.3°C, 17.5°C)	6
Cluster (22.9°C, 22,2°C)	7
Cluster (4:22 p.m., 4:16 p.m., 4:19 p.m., 4:26 p.m.)	8

c, create temporal database with item IDs

Date		08/18				08/19				08/20				08/21			
Time		6:39 a.m.	8:30 a.m.	4:22 p.m.	5:21 p.m.	7:24 a.m.	8:15 a.m.	4:16 p.m.	8:14 p.m.	8:05 a.m.	8:25 a.m.	4:19 p.m.	10:09 p.m.	6:33 a.m.	8:33 a.m.	4:26 p.m.	9:36 p.m.
Event		Win open	Door open	Door close	Win close	Win open	Door open	Door close	Win close	Win open	Door open	Door close	Win close	Win open	Door open	Door close	Win close
Temporal database	Item ID (event)	1	3	4	2	1	3	4	2	1	3	4	2	1	3	4	2
	Item ID (ambient)	5		8	6	5		8	6	5		8	7			8	7

Figure 3. Convert sensor data tables into temporal database

event is assigned with ID 1. This will be recorded in the table b item ID mapping table.

Second, get all the associated ambient data for each event, and try to cluster with DBSCAN algorithm. In table a, there are two kinds of ambient data: time and temperature. For example, in sub-table 2, the window-opening event occurs at 6:39 A.M., 7:24 A.M., 8:05 A.M., and 6:33 A.M. on each day. In sub-table 3, we can get the temperature at these time points, which are 27.5°C, 26.8°C, 28.4°C, and 22.5°C. By clustering these data with DBSCAN and 1°C as max distance, the first three temperatures, 27.5 °C, 26.8 °C, 28.4 °C, can be clustered into one cluster, while the fourth one, 22.5 °C, will be defined as noise. No cluster can be found for the window-opening and closing event with 15 minutes distance. The new found temperature cluster is also be assigned with an item ID 5.

After all associated ambient data cluster are found and assigned with item IDs, all events will be sorted in an array by its timestamp. Their associated ambient data will also be added into the same itemset (in table c). In this case, the temporal database will be like this: ⟨(1 5)(3)(4 8)(2 6)(1 5)(3)(4 8)(2 6)(1 5)(3)(4 8)(2 7)(1)(3)(4 8)(2 7)⟩ .

C. Frequent episode mining module

Although FEM derives from FPM and shares a lot of similarities, there are still some differences in the algorithm. One essential difference in mining the temporal database and sequence database is the frequency metric, i.e. the amount of occurrence of a pattern. However, no measures have been commonly accepted in temporal database mining. In this paper, a metric called LMaxnR-freq [11], which is short for the Leftmost Maximal non-Redundant set of occurrences, is adopted.

In the mining process, the length of each frequent episode grows by iteration. Each length L frequent episode search for L+1 episode in the projected database with the metric defined in LMaxnR-freq. All the found episodes are maintained in a tree structure called enumeration tree. There, episodes will be pruned or kept for the next iteration. The actual implement of the algorithm follows these papers [12, 13].

Take the case in Figure 3 for example, with a min_sup = 3, one of the frequent episode, ⟨(1 5)(3)(4 8)(2)⟩ , can be found. This episode can be translated into the behavior pattern: in the morning the window will be opened when the temperature is around 26.5 °C and then the door will be opened, around 4:20 P.M. the door will be closed and afterward also the window.

D. Visualization module

The visualization module visualizes the data created during the whole mining process. Unlike the most of the algorithms just provides the frequent episodes and their supports, with this visualization module, researchers are able to track occurrences of the patterns and understand actual meanings of the patterns. There are three different charts: sensor data table chart, enumeration tree chart, and frequent episode chart.

The sensor data table chart displays the content of the table, the metadata from the sensor metadata table, and some statistics of the table, such as the data type of the sensor value, the amount of different values and the size of the table, etc.

The enumeration tree chart and frequent episode chart work together to display the information of the frequent episode. In enumeration tree chart, each circle represents a node in the tree structure. The different color represents the different state of the node: black is pruned and, yellow is not pruned. The number, in the circle, on the left side of the colon is the item ID of the node, right side is the frequency of the episode. The link between the nodes represents the type of extension: solid line is horizontal extension and dash line is the vertical extension. The frequent episode chart is a 2D-grid, each row contains the occurrences of a certain item ID, and each column represents a timestamp in the temporal database. When a certain node is selected in the enumeration tree chart, the corresponding occurrences of this episode will be marked in the frequent episode chart. Each item in one occurrence will be connected by red line (Figure 4).

V. TEST OF THE FRAMEWORK

A. The framework of the pattern discovery process

The framework consists of three parts: data acquisition, data managing, and data mining (Figure 5). The data acquisition task is done by the WSN, which contains both active sensor and passive sensor node. The ambient sensor periodically records the environment parameters. The event sensor records the events triggered by occupants. Data are stored and maintained in the form of tables in database. There are event sensor data and ambient sensor data for different recording. The metadata table provides the spatial relationship for linking the event sensor tables with ambient sensor tables. With the data and their relationships in the database, the temporal database can be created for mining. Then the WSN data can be converted into a normal FEM problem. After mining process, the discovered frequent episodes can be translated into behavior patterns with the information in the database.

There are several facts that can affect the output patterns. The first is the features of input sensor data, including the density of the sequential pattern, and the average length of the patterns. The second is the parameters for clustering the ambient sensor data, i.e., the max

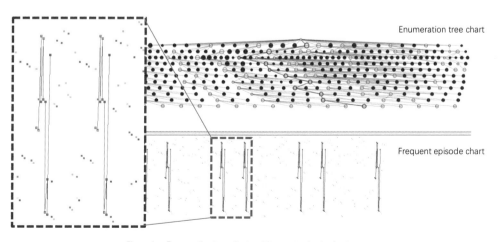

Figure 4. Enumeration tree chart and frequent episode chart

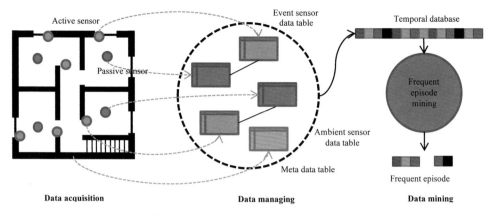

Figure 5. Diagram for pattern discovery process

distance between each data points. The third is the parameters for the FEM mining algorithm, including the minimum support of the pattern (min_sup), the size of the window constrain (max_gap), and the degree of approximation for pruning the sub-sequence (max_err_bound). However, these parameters can be decided by evaluating the data source without additional knowledge.

Two tests have been conducted: the first one is based on the synthetic data generated by a simulator, which can generate sensor data records mix with patterns and noises. The second test is based on the real life data collected from a student dormitory, which provides a demonstration and evaluation of the proposed framework in real world application.

B. Experiment with synthetic data

In this experiment, one long daily pattern is predefined. The virtual environment generated 10 days sensor data with 10% noise data, 320 data samples were collected. 48 items were generated to represent all events and associated ambient data groups. With minimum support set to 10, the mining process lasted for 93 iterations. In the final iteration, 14026 tree nodes were found, and 14019 of them were pruned. Only 7 nodes were left. They were:

1, <18 15 >:29
2, <36 33 >:19
3, <24 30 27 21 >:19
4, < (18 19) (15 16) >:20
5, <18 (4 15) 1 >:19
6, <46 43 >:19
7, <(24 25) (30 31) (27 28) (18 19 21 22) (4 5 15 16) (1 2 18 19) (15 16 46 47) (36 37 43 44) (33 34) (36 38) (41 42) (33 35 46 48) (18 20 43 45) (4 6 15 17) (9 10) (1 3 13 14) (11 12 24 26) (30 32) (27 29) (21 23) (24 25) (30 31) (27 28) (18 19 21 22) (4 5 15 16) (7 8) (1 2 18 19) (15 16 46 47) (36 37 43 44) (39 40) (33 34) >:9

Figure 6-a shows the occurrences of the seventh pattern. It is a length-93 episode and covers the whole daily routine as expected. This experiment shows, the algorithm is able to detect all of the predefined patter with noise.

C. Experiment with real life data

In this experiment, data collected with the self-organizing WSN installed in a room of stu-

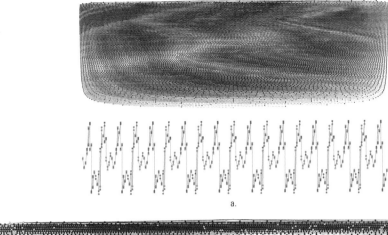

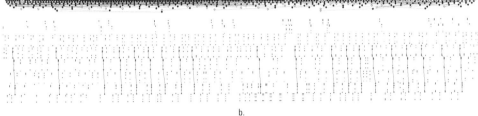

Figure 6. a. Enumeration tree and occurance of pattern 7 with synthetic data,
b. Enumeration tree and patter <(25 28 34) (36 37 45) (59 61 67) > with real life data

dent dormitory for a one month period is used for testing.

The test environment was a 12 m^2 single room. A self-organizing WSN with eight nodes was installed in the room, including one data logging node, five passive sensor nodes, and two active sensor nodes. Two of the passive sensor nodes were placed outdoors to monitor the outdoor, and corridor environment, respectively. The remaining three were placed next to the desk, bed, and basin, respectively. The active sensor nodes were respectively installed on the window frame and door frame to monitor their opening and closing state.

With 30 days period, 732 data sample have been recorded. 75 items were generated. The min_sup was set to 20 (there were 20 working days), the constrain window size was set to 3, and the max_error_bound was set to 5. The searching ended up with 13 iterations. There were 1613 nodes in the enumeration tree (Figure 6-b), 1450 of them were pruned, and 163 were left.

Frequent episodes with associated ambient value clusters can be translated into behavior patterns. For example, the pattern <(25 28 34) (36 37 45) (59 61 67) >:20 (Figure 14) can be interpreted as: desk PIR sensor was activated around 6:35 with room light on, bed PIR sensor deactivated around 6:45 with room light, lamp turned off around 7:15, room light off. This was a recode of the period when the occupant got up in the morning and turned off the lamp before leaving the room. The desk PIR sensor activated before the bed PIR sensor, because there was a delay before PIR sensor deactivated.

VI. CONCLUSION AND FUTURE WORK

As a conclusion, the proposed framework is able to collect sensor data from building and discover behavior patterns. The advantage of the proposed framework is: (1) It is easy to deploy. (2) It less data collection and calculation. (3) It needs very little amount of settings and parameters. (4) It can work with both numerical and categorical data, and the output pattern contains both sensor event and corresponding ambient values. (5) We also first visualized the mining processing and result.

There are some possible future works. First, in the real life data test, there are still redundancy existed. It is caused by the parallel patterns, in which some items shifted in sequence. By introducing parallel episode, the output can be more condense. Second, the framework is not limited for building sensor, it can be easily extended to discovery people's daily routine in city by add the mobile location data, and etc.

REFERENCES

[1] Xu, K., et al., Toward software defined smart home. IEEE Communications Magazine, 2016. 54(5): p. 116-122.

[2] Mehmood, A. and H. Song, Smart Energy Efficient Hierarchical Data Gathering Protocols for Wireless Sensor Networks. Smart Computing Review, 2015. 5(5): p. 425-462.

[3] Ahmadi, A., et al., An efficient routing algorithm to preserve k -coverage in wireless sensor networks. The Journal of Supercomputing, 2014. 68(2): p. 599-623.

[4] Shojafar, M., N. Cordeschi, and E. Baccarelli, Energy-efficient Adaptive Resource Management for Real-time Vehicular Cloud Services. IEEE Transactions on Cloud Computing, 2016. PP(99): p. 1-1.

[5] Baek, J., et al., On a moving direction pattern based MAP selection model for HMIPv6 networks. Computer Communications, 2011. 34(2): p. 150-158.

[6] Sun, Y., Q. Jiang, and M. Singhal, A Pre-Processed Cross Link Detection Protocol for geographic routing in mobile ad hoc and sensor networks under realistic environments with obstacles. Journal of Parallel and Distributed Computing, 2011. 71(7): p. 1047-1054.

[7] Rashid, M.M., I. Gondal, and J. Kamruzzaman, Dependable large scale behavioral patterns mining from sensor data using Hadoop platform. Information Sciences, 2016.

[8] Agrawal, R., et al., Mining association rules between sets of items in large databases. SIGMOD Rec., 1993. 22(2): p. 207-216.

[9] Gan, M. and H. Dai, Fast Mining of Non-derivable Episode Rules in Complex Sequences, in Modeling Decision for Artificial Intelligence, V. Torra, et al., Editors. 2011, Springer Berlin Heidelberg. p. 67-78.

[10] Ester, M., et al. A density-based algorithm for discovering clusters in large spatial databases with noise. in Kdd. 1996.

[11] Gan, M. and H. Dai, Subsequence Frequency Measurement and its Impact on Reliability of Knowledge Discovery in Single Sequences, in Reliable Knowledge Discovery, H. Dai, J.N.K. Liu, and E. Smirnov, Editors. 2012, Springer US. p. 239-255.

[12] Han, J., et al., Mining Frequent Patterns without Candidate Generation: A Frequent-Pattern Tree Approach. Data Mining and Knowledge Discovery, 2004. 8(1): p. 53-87.

[13] Min, G. and D. Honghua. A Study on the Accuracy of Frequency Measures and Its Impact on Knowledge Discovery in Single Sequences. in Data Mining Workshops (ICDMW), 2010 IEEE International Conference on. 20.

12

高精度多目标实时定位及分析系统
High-Precision Multi-Targets Real-Time Locating and Analyzing System

设计者：李力，Nexiot
Designer: Li Li , Nexiot

人在建筑中的行为模式对建筑设计和建筑节能都有重要影响。然而，目前受测量精度的限制，还没有成熟的观测手段。Inst. AAA 和 Nexiot 联合研制的室内定位系统利用超宽带无线定位和双向测距技术，将定位精度提升到厘米级。同时，利用优化算法，使得在跟踪大量目标时，还能保持较高的更新频率。此外，系统还集成了实时监控和时空分布热力图计算功能，为研究人员提供更直观的结果。

该系统已有很多成功的应用案例。例如，在瑞士最大零售商的顾客购物行为分析项目中，该系统就被用来记录顾客在大型超市中的购物流线。监测过程中，8 个定位基站覆盖了 1200m^2 的货架空间，定位标签放置在购物篮和购物车上。经过一周的数据积累，不同人群的购物行为特征和兴趣点很容易就能在图表中体现出来。

Occupant's behavior pattern is a very influential factor in architectural design and energy conservation. However, due to the limitation in precision, there is still no mature monitoring technology. The RTLS developed by Inst.AAA and Nexiot overcomes these limitations. By utilizing Ultra-wideband and two-way ranging techniques, the precision reaches centimeter level. With optimizing algorithm, large amount of targets can be tracked while still keeping high update frequency. Meanwhile, real time monitoring interface and heat map generation are also integrated to provide researchers with more intuitive results.

There are already several successful applications. For example, in the customer shopping behavior analysis project for one of the largest Swiss retailers, this system has been used for tracking their customers' circulation in a supermarket. Eight base stations covers 1200 m2 space and tracking devices are installed on the shopping basket. After one week of monitoring, points of interest and behavior pattern could be easily visualized.

锚点与标签（Anchro & Tag）

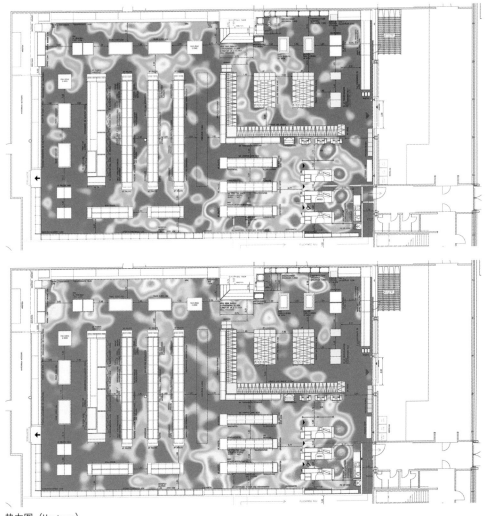

热力图(Heatmap)

后记

十年磨一剑，寒刃未曾试。

这本书是东南大学建筑学院建筑运算与应用研究所（Inst. AAA）十年来的教学与研究记录。整本书的编撰过程中我努力去读懂那些未曾参与过的项目，理解它们在"铸剑"过程中的定位和意义。尤其是一些早期的案例：数控建造"Angle-X"，互动设计"Lightbox"等，现在看来司空见惯的数字技术的尝试，当年需要在没有多少人理解的环境中持续不断地探索和争取方能做成。而正是这些作品铸成了这把剑的雏形，在此基础上的砥砺前行，成就了这本书里的大部分内容。近期的一些作品，"映沙"、"Neuron"等，其意思也不只在于其对感官的高度刺激，更重要的是设计者对建筑问题变革性的想法。

数字技术虽是新生事物，但其重技术求创新的本质不允许科研团队停停走走。不断地发现与探索的需求，使得这个领域里真正的勇者为数不多。这本书只是一个记录，是高山流水觅知音，不算宣言更不为荣耀。

尘嚣远遁，铅华独守，新书出版之际，建筑运算与应用研究所的同仁们已经在新的领域继续踽踽前行。

唐芃
2017 年 9 月
于东南大学前工院

POSTSCRIPT

Ten years for grinding a sword, frost blade never tried.

This book is a record of ten years of teaching and research of the Institute of Architectural Algorithms and Applications (Inst. AAA) of the School of Architecture, Southeast University. During the compilation of the book, I tried to understand the projects that I haven't been involved and understand their positions and meanings in the process of "swording". Especially some early cases as: digital fabrication "Angle-X", interactive design "Lightbox" and so on, many attempts which now look familiar with the digital technology, but called for continuous exploring and striving to realize, in a less understood environment in those days. Just these pieces of works cast into the embryonic form of this "sword", most of the contents of the book are the achievements of the diligence and courage forward on the basis. Recent works like "Sand Mapping", "Neuron" and so on, are not only expressing intense sensory stimuli, more importantly, representing the designer's transformational ideas about the architectural problems.

Digital technology is a new thing, but the essence of focusing on technological innovation doesn't allow the research team to stop-and-go. Demand of constantly discovering and exploring makes the real brave not much. This book is just a record to seek some bosom friend, not a declaration, less for honor.

When the book published, colleagues of Inst. AAA have already step on a new journey ahead.

Tang Peng
September 2017
In Qiangongyuan, Southeast University